Lecture Notes in Mathematics

Edited by A. Dold and B. Eckmann

1279

Dorin Ieşan

Saint-Venant's Problem

Springer-Verlag

Berlin Heidelberg New York London Paris Tokyo

Author

Dorin Ieşan
Department of Mathematics, University of Iaşi
6600 Iaşi, Rumania

Mathematics Subject Classification (1980): 73C10, 73C20, 73C25, 73C30, 73C40, 73K05, 73B25, 35J55

ISBN 3-540-18361-2 Springer-Verlag Berlin Heidelberg New York
ISBN 0-387-18361-2 Springer-Verlag New York Berlin Heidelberg

Library of Congress Cataloging-in-Publication Data. Ieşan, Dorin. Saint-Venant's problem. (Lecture notes in mathematics; 1279) Bibliography: p. Includes index. 1. Saint-Venant's principle. 2. Elasticity. 3. Cylinders. I. Title. II. Series: Lecture notes in mathematics (Springer-Verlag; 1279. QA3.L28 no. 1279 [QA931] 510 s 87-23553 ISBN 0-387-18361-2 (U.S.) [624.1'772]

Printing and binding: Druckhaus Beltz, Hemsbach/Bergstr.
2146/3140-543210

INTRODUCTION

A major concern throughout the history of elasticity has been with problems dictated by the demands of engineering. Interest in the construction of a theory for the deformation of elastic cylinders dates back to Coulomb, Navier and Cauchy. However, only Saint-Venant has been able to give a solution of the problem.

The importance of Saint-Venant's celebrated memoirs [132,133], on what has long since become known as Saint-Venant's problem requires no emphasis. To review the vast literature to which the work contained in [132,133] has given impetus is not our intention. An account of the historical developments as well as references to various contributions, may be found in the books and in some of the papers cited.

We recall that Saint-Venant's problem consists in determining the equilibrium of a homogeneous and isotropic linearly elastic cylinder, loaded by surface forces distributed over its plane ends. Saint-Venant proposed an approximation to the solution of the three-dimensional problem, which only requires the solution of two-dimensional problems in the cross section of the cylinder. Saint-Venant's formulation leads to the four basic problems of extension, bending, torsion and flexure. His analysis is founded on physical intuition and elementary beam theory. Saint-Venant's extension, bending, torsion, and flexure solutions are well-known (see, for example, Love [98], Chapters 14,15 and Sokolnikoff [139], Chapter 4).

Saint-Venant's approach of the problem is based on a relaxed statement in which the pointwise assignment of the terminal tractions is replaced by prescribing the corresponding resultant force and resultant moment. Justification of the procedure is twofold. First, it is difficult in practice to determine the actual distribution of applied stresses on the ends, although the resultant force and moment can be measured accurately. Second, one invokes Saint-Venant's principle. This principle states, roughly speaking, that if two sets of loadings are statically equivalent at each end, then the difference in stress fields and strain fields are negligible, except possibly near the ends. The precise meaning of Saint-Venant's hypothesis and its justification have been the subject of many studies, almost from the time of the original Saint-Venant's papers. Reference to some of the early investigations of the question will be found in [98],[139],[140]. In recent years important steps toward clarifying Saint-Venant's principle have been made. The classic paper in linear elasticity is by Toupin [146] (see also, e.g. Roseman [128], Knowles [84] and Fichera [39,40] for further important developments). For the history of the problem and the detailed analysis of various results on Saint-Venant's principle we refer to the works of M.E.Gurtin [47], G.Fichera [38], C.O.Horgan and J.K.Knowles [53].

The relaxed Saint-Venant's problem continues to attract attention both from the mathematical and the technical point of view.

It is obvious that the relaxed statement of the problem fails to characterize the solution uniquely. This fact led various authors to establish characterizations of Saint-Venant's solution. Thus, Clebsch [24] proved that Saint-Venant's solution can be derived from the assumption that the stress vector on any plane normal to the cross-sections of the cylinder is parallel to its generators. In [155], Voigt rediscovered Saint-Venant's solution by using another assumption regarding the structure of the stress field. Thus, Saint-Venant's extension, bending and torsion solutions are derived from the hypothesis that the stress field is independent of the axial coordinate, and Saint-Venant's flexure solution is obtained if the stress field depends on the axial coordinate at most linearly.

E.Sternberg and J.K.Knowles [143] characterized Saint-Venant's solutions in terms of certain associated minimum strain-energy properties. Other intrinsic criteria that distinguish Saint-Venant's solutions among all the solutions of the relaxed problem were established in [79]. In [79], a rational scheme of deriving Saint-Venant's solutions is presented. The advantage of this method is that it does not involve artificial a priori assumptions. The method permits to construct a solution of the relaxed Saint-Venant's problem for other kinds of constitutive equations (anisotropic media, Cosserat continua, etc.) where the physical intuition or semi-inverse method cannot be used.

In [148]-[150], C.Truesdell proposed a problem which, roughly speaking, consists in the generalization of Saint-Venant's notion of twist so as to apply to any solution of the torsion problem. Recently an elegant solution of Truesdell's problem has been established by W.A.Day [25]. In [123], P.Podio-Guidugli studied Truesdell's problem rephrased for extension and pure bending. The case of flexure was considered in [79]. The results of [25,123] are related to the results of Sternberg and Knowles [143] concerning the minimum energy characterizations of corresponding Saint-Venant's solutions.

A generalization of the relaxed Saint-Venant's problem consists in determining the equilibrium of an elastic cylinder which - in the presence of body forces - is subjected to surface tractions arbitrarily prescribed over the lateral boundary and to appropriate stress resultants over its ends. The study of this problem was initiated by Almansi [1] and Michell [102] and was developed in various later papers (see,for example, Sokolnikoff [139], Djanelidze [29] and Hatiashvili [49]).

As pointed out before, Saint-Venant's results were established within the equilibrium theory of homogeneous and isotropic elastic bodies. A large number of papers are concerned with the relaxed Saint-Venant's problem for other kinds of elastic materials (see, for example, Lekhnitskii [96], Lomakin [97], Brulin and R.K.T.Hsieh [15] and Reddy and Venkatasubramanian [90]). References to recent results are cited in the text. No attempt is made to provide a complete list of works on Saint-Venant's problem. Neither the contents, nor the list of works cited are exhaustive. Nevertheless, it is hoped that the developments presented reflect the state of knowledge in the study of the problem.

The purpose of this work is to present some of the recent researches on Saint-Venant's problem. An effort is made to provide a systematic treatment of the subject.

Chapter 1 is concerned mainly with results where Saint-Venant's solutions are involved. We give a rational method of construction of these solutions and then we characterize them in terms of certain associated minimum strain-energy properties. A study of Truesdell's problem is presented. This chapter also includes a proof of Saint-Venant's principle.

In Chapter 2, an interesting scheme of deriving a solution of

Almansi-Michell problem is presented. Almansi's problem, where the body forces and the surface tractions on the lateral boundary are polynomials in the axial coordinate, is also studied. The results are used to study a statical problem in the linear thermoelasticity.

Chapter 3 is concerned with the relaxed Saint-Venant's problem for anisotropic elastic bodies. We first establish a solution of the foregoing problem. The method does not involve artificial a priori assumptions and permits a treatment of the problem even for nonhomogeneous bodies, where the elastic coefficients are independent of the axial coordinate. It is shown that the well-known boundary-value problem for the torsion function derives from a special problem of generalized plane strain. Then, minimum energy characterizations of the solutions are presented. Also included in this chapter is a study of Truesdell's problem.

Chapter 4 deals with the relaxed Saint-Venant's problem for heterogeneous elastic cylinders. We consider the case of a composed cylinder when the generic cross-section is occupied by different anisotropic solids. The problems of Almansi and Michell are also studied. Applications to the linear thermoelastostatics are given.

In Chapter 5 we study Saint-Venant's problem within the linearized theory of Cosserat elastic bodies. We first present a proof of Saint-Venant's principle in the theory of Cosserat elasticity. Then, a solution of the relaxed Saint-Venant's problem is derived. Truesdell's problem and a theory of loaded cylinders are also studied. Illustrative applications are presented.

A number of results included in this work have not appeared or been discussed previously in literature.

CONTENTS

1. THE RELAXED SAINT-VENANT PROBLEM

1.1. Preliminaries

We refer to a bounded regular region B of three-dimensional Euclidian space as the body (cf. M.E. Gurtin [47], Sect. 5). We let $\bar{B}$ denote the closure of B, call ∂B the boundary of B, and designate by $\underline{n}$ the outward unit normal of ∂B. Letters marked by an underbar stand for tensors of an order $p \geq 1$, and if $\underline{v}$ has the order p, we write $v_{ij\ldots k}$ (p subscripts) for the components of $\underline{v}$ in the underlying rectangular Cartesian coordinate frame. We shall employ the usual summation and differentiation conventions: Greek subscripts are understood to range over the integers (1,2), whereas Latin subscripts - unless otherwise specified - are confined to the range (1,2,3); summation over repeated subscripts is implied and subscripts preceded by a comma denote partial differentiation with respect to the corresponding Cartesian coordinate.

We assume that the body occupying B is a linearly elastic material. Let $\underline{u}$ be a displacement field over B. Then

$$\underline{E}(\underline{u}) = \operatorname{sym} \nabla \underline{u} \ ,$$

is the strain field associated with $\underline{u}$. Here $\nabla \underline{u}$ denotes the displacement gradient and $(\operatorname{sym} \nabla \underline{u})_{ij} = (u_{i,j} + u_{j,i})/2$. The stress-displacement relation may be written in the form

$$\underline{S}(\underline{u}) = \underline{C}[\nabla \underline{u}]. \qquad (1.1)$$

Here $\underline{S}(\underline{u})$ is the stress field associated with $\underline{u}$, while $\underline{C}$ stands for the elasticity field. We assume that $\underline{C}$ is positive-definite, symmetric, and smooth on $\bar{B}$. For the particular case of the isotropic elastic medium the tensor field $\underline{C}$ admits the representation

$$C_{ijk\ell} = \lambda \delta_{ij}\delta_{k\ell} + \mu (\delta_{ik}\delta_{j\ell} + \delta_{i\ell}\delta_{jk}), \qquad (1.2)$$

where λ and μ are the Lamé moduli and δ_{ij} is the Kronecker delta.

We call a vector field $\underline{u}$ an equilibrium displacement field for B if $\underline{u} \in C^1(\bar{B}) \cap C^2(B)$ and

$$\operatorname{div} \underline{S}(\underline{u}) = \underline{0} \ , \tag{1.3}$$

holds on B. Clearly, $(\operatorname{div} \underline{S}(\underline{u}))_{ij} = (S_{ij}(\underline{u}))_{,j}$.

Let $\underline{s}(\underline{u})$ be the surface traction at regular points of ∂B corresponding to the stress field $\underline{S}(\underline{u})$ defined on $\bar{B}$, i.e.

$$\underline{s}(\underline{u}) = \underline{S}(\underline{u})\underline{n} \ . \tag{1.4}$$

The strain energy $U(\underline{u})$ corresponding to a smooth displacement field $\underline{u}$ on $\bar{B}$ is (cf.[47], Sect.32)

$$U(\underline{u}) = \frac{1}{2}\int_B \nabla \underline{u} \cdot \underline{C}[\nabla \underline{u}]dv \ . \tag{1.5}$$

In the following, two displacement fields differing by an (infinitesimal) rigid displacement will be regarded identical.

The functional $U(\cdot)$ generates the bilinear functional

$$U(\underline{u},\underline{v}) = \frac{1}{2}\int_B \nabla \underline{u} \cdot \underline{C}[\nabla \underline{v}]dv \ .$$

The set of smooth fields over $\bar{B}$ can be made into a real vector space with the inner product

$$\langle \underline{u},\underline{v} \rangle = 2\, U(\underline{u},\underline{v}) \ . \tag{1.6}$$

This inner product generates the energy norm

$$\| \underline{u} \|_e^2 = \langle \underline{u},\underline{u} \rangle \ . \tag{1.7}$$

For any equilibrium displacement fields $\underline{u}$ and $\underline{v}$, one has (cf. [47], Sect.30)

$$\langle \underline{u},\underline{v} \rangle = \int_{\partial B} \underline{s}(\underline{v})\cdot\underline{u}\, da \ , \tag{1.8}$$

which implies the well known relation

$$\int_{\partial B} \underline{s}(\underline{u})\cdot\underline{v}\, da = \int_{\partial B} \underline{s}(\underline{v})\cdot\underline{u}\, da \ . \tag{1.9}$$

1.2. Properties of the Solutions to the Relaxed Saint-Venant Problem

We proceed now to Saint-Venant's problem and for this purpose stipulate that the region B from here on refers to the interior of a right cylinder of length h with open cross-section Σ and the lateral boundary Π. The rectangular Cartesian coordinate frame is supposed to be chosen in such a way that the x_3-axis is parallel to the generators of B and the x_1Ox_2 plane contains one of the terminal cross-sections. We denote by Σ_1 and Σ_2, respectively, the cross-section located at $x_3 = 0$ and $x_3 = h$. We assume that the generic cross-section Σ is a simply connected regular region. We denote by Γ the boundary of Σ.

Saint-Venant's problem consists in the determination of an equilibrium displacement field $\underline{u}$ on B, subject to the requirements

$$\underline{s}(\underline{u}) = \underline{0} \quad \text{on} \quad \Pi, \qquad \underline{s}(\underline{u}) = \underline{s}^{(\alpha)} \quad \text{on} \quad \Sigma_\alpha (\alpha = 1,2) , \tag{1.10}$$

where $\underline{s}^{(\alpha)}$ is a vector-valued function preassigned on Σ_α. Necessary conditions for the existence of a solution to this problem are given by

$$\int_{\Sigma_1} \underline{s}^{(1)} da + \int_{\Sigma_2} \underline{s}^{(2)} da = \underline{0} , \qquad \int_{\Sigma_1} \underline{x} \times \underline{s}^{(1)} da + \int_{\Sigma_2} \underline{x} \times \underline{s}^{(2)} da = \underline{0} ,$$

where $\underline{x}$ is the position vector of a point with respect to O.

Under suitable smoothness hypotheses on Γ and on the given forces, a solution of Saint-Venant's problem exists (cf. Fichera [37]).

In the relaxed formulation of Saint-Venant's problem the conditions (1.10) are replaced by

$$\underline{s}(\underline{u}) = \underline{0} \quad \text{on} \quad \Pi , \quad \underline{R}(\underline{u}) = \underline{F} , \quad \underline{H}(\underline{u}) = \underline{M} , \tag{1.11}$$

where $\underline{F}$ and $\underline{M}$ are prescribed vectors representing the resultant force and the resultant moment about O of the tractions acting on Σ_1. Accordingly, $\underline{R}(\cdot)$ and $\underline{H}(\cdot)$ are the vector-valued linear functionals defined by

$$\underline{R}(\underline{u}) = \int_{\Sigma_1} \underline{s}(\underline{u}) da, \qquad \underline{H}(\underline{u}) = \int_{\Sigma_1} \underline{x} \times \underline{s}(\underline{u}) da . \tag{1.12}$$

If $\varepsilon_{\alpha\beta}$ is the two-dimensional alternator, (1.12) appears as

$$R_i(\underline{u}) = -\int_{\Sigma_1} S_{3i}(\underline{u})da,$$

$$H_\alpha(\underline{u}) = -\int_{\Sigma_1} \varepsilon_{\alpha\beta} x_\beta S_{33}(\underline{u})da, \quad H_3(\underline{u}) = -\int_{\Sigma_1} \varepsilon_{\alpha\beta} x_\alpha S_{3\beta}(\underline{u})da \,. \tag{1.13}$$

The necessary conditions for the existence of a solution to Saint-Venant problem lead to the following relations, which are needed subsequently

$$\int_{\Sigma_2} S_{3i}(\underline{u})da = -R_i(\underline{u}) \,, \int_{\Sigma_2} \varepsilon_{\alpha\beta} x_\alpha S_{3\beta}(\underline{u})da = -H_3(\underline{u}),$$

$$\int_{\Sigma_2} x_\alpha S_{33}(\underline{u})da = -hR_\alpha(\underline{u}) + \varepsilon_{\alpha\beta} H_\beta(\underline{u}) \,. \tag{1.14}$$

It is obvious that the relaxed statement of the problem fails to characterize the solution uniquely.

By a solution of the relaxed Saint-Venant's problem we mean any equilibrium displacement field that satisfies (1.11).

We denote by (P) the relaxed Saint-Venant's problem corresponding to the resultants $\underline{F}$ and $\underline{M}$. Let $K(\underline{F},\underline{M})$ denote the class of solutions to the problem (P).

The classification of the relaxed problem rests on various assumptions concerning the resultants $\underline{F}$ and $\underline{M}$. Throughout this work it is convenient to use the decomposition of the relaxed problem into problems (P_1) and (P_2) characterized by

(P_1) (extension-bending-torsion) : $F_\alpha = 0$,

(P_2) (flexure) : $F_3 = M_i = 0$.

For further economy it is helpful to denote by $K_I(F_3,M_1,M_2,M_3)$ the class of solutions to the problem (P_1) and by $K_{II}(F_1,F_2)$ the class of solutions to the problem (P_2). We assume for the remainder of this chapter that the material is homogeneous and isotropic.

Let $\mathcal{D}$ denote the set of all equilibrium displacement fields $\underline{u}$ that satisfy the condition $\underline{s}(\underline{u}) = \underline{0}$ on the lateral boundary. The next theorem will be of future use.

Theorem 1.1. [79]. If $\underline{u} \in \mathcal{D}$ and $\underline{u}_{,3} \in C^1(\bar{B})$, then $\underline{u}_{,3} \in \mathcal{D}$ and

$$\underline{R}(\underline{u}_{,3}) = 0 \ , \quad H_\alpha(\underline{u}_{,3}) = \varepsilon_{\alpha\beta} R_\beta(\underline{u}) \ , \quad H_3(\underline{u}_{,3}) = 0. \tag{1.15}$$

Proof. The first assertion follows at once from the fact that $\underline{S}(\underline{u}_{,3}) = \partial \underline{S}(\underline{u})/\partial x_3$ and the proposition: if $\underline{u}$ is an elastic displacement field corresponding to null body forces, then so also is $\underline{u}_{,k} = \partial \underline{u}/\partial x_k$ (cf.[47],Sect.42). Next, with the aid of the equations of equilibrium (1.3), we find that

$$S_{3i}(\underline{u}_{,3}) = (S_{3i}(\underline{u}))_{,3} = -(S_{\rho i}(\underline{u}))_{,\rho} \ ,$$

$$\varepsilon_{\alpha\beta} x_\beta S_{33}(\underline{u}_{,3}) = -\varepsilon_{\alpha\beta} x_\beta (S_{\rho 3}(\underline{u}))_{,\rho} = -\varepsilon_{\alpha\beta}[(x_\beta S_{\rho 3}(\underline{u}))_{,\rho} - S_{\beta 3}(\underline{u})] \ ,$$

$$\varepsilon_{\alpha\beta} x_\alpha S_{3\beta}(\underline{u}_{,3}) = -\varepsilon_{\alpha\beta} x_\alpha (S_{\rho\beta}(\underline{u}))_{,\rho} = -\varepsilon_{\alpha\beta}(x_\alpha S_{\rho\beta}(\underline{u}))_{,\rho} + \varepsilon_{\alpha\beta} S_{\alpha\beta}(\underline{u}).$$

By (1.13), the divergence theorem, and the symmetry of $\underline{S}$ we arrive at

$$\begin{aligned} \underline{R}(\underline{u}_{,3}) &= \int_\Gamma \underline{s}(\underline{u})\,ds, \\ H_\alpha(\underline{u}_{,3}) &= \int_\Gamma \varepsilon_{\alpha\beta} x_\beta s_3(\underline{u})\,ds + \varepsilon_{\alpha\beta} R_\beta(\underline{u}), \\ H_3(\underline{u}_{,3}) &= \int_\Gamma \varepsilon_{\alpha\beta} x_\alpha s_\beta(\underline{u})\,ds \ . \end{aligned} \tag{1.16}$$

The desired result follows from (1.16) and hypothesis. □

Since $\underline{u}$ is an equilibrium displacement field, $\underline{u}$ is analytic (cf. [47], Sect.42). Theorem 1.1 has the following immediate consequences:

Corollary 1.1. If $\underline{u} \in K_I(F_3, M_1, M_2, M_3)$ and $\underline{u}_{,3} \in C^1(\bar{B})$, then $\underline{u}_{,3} \in \mathcal{D}$ and

$$\underline{R}(\underline{u}_{,3}) = \underline{0} \ , \quad \underline{H}(\underline{u}_{,3}) = \underline{0} \ .$$

Corollary 1.2. If $\underline{u} \in K_{II}(F_1, F_2)$ and $\underline{u}_{,3} \in C^1(\bar{B})$, then

$$\underline{u}_{,3} \in K_I(0, F_2, -F_1, 0).$$

Corollary 1.3. If $\underline{u} \in \mathcal{D}$ and $\partial^n \underline{u}/\partial x_3^n \in C^1(\bar{B})$, then $\partial^n \underline{u}/\partial x_3^n \in \mathcal{D}$ and

$$\underline{R}\left(\frac{\partial^n \underline{u}}{\partial x_3^n}\right) = \underline{0}\ , \quad \underline{H}\left(\frac{\partial^n \underline{u}}{\partial x_3^n}\right) = \underline{0} \quad \text{for } n \geq 2.$$

1.3. A Method of Construction of Saint-Venant's Solutions

Corollary 1.1 allows us to establish a simple method of deriving Saint-Venant's solution to the problem (P_1). Let Q be the class of solutions to the relaxed Saint-Venant's problem corresponding to $\underline{F} = \underline{0}$ and $\underline{M} = \underline{0}$. We note that if $\underline{u} \in K_I(F_3, M_1, M_2, M_3)$ and $\underline{u}_{,3} \in C^1(\bar{B})$, then by Corollary 2.1, $\underline{u}_{,3} \in Q$. Let us note that a rigid displacement field belongs to Q. It is natural to enquire whether there exists a solution $\underline{v}$ of the problem (P_1) such that $\underline{v}_{,3}$ is a rigid displacement field. This question is settled in the next theorem.

Theorem 1.2. Let $\underline{v} \in C^1(\bar{B}) \cap C^2(B)$ be a vector field such that $\underline{v}_{,3}$ is a rigid displacement field. Then $\underline{v}$ is a solution of the problem (P_1) if and only if $\underline{v}$ is Saint-Venant's solution.

Proof. Let $\underline{v} \in C^1(\bar{B}) \cap C^2(B)$ be a vector field such that

$$\underline{v}_{,3} = \underline{\alpha} + \underline{\beta} \times \underline{x}\ , \tag{1.17}$$

where $\underline{\alpha}$ and $\underline{\beta}$ are constant vectors. Then it follows that

$$\begin{aligned} v_\alpha &= -\tfrac{1}{2} a_\alpha x_3^2 - a_4 \varepsilon_{\alpha\beta} x_\beta x_3 + w_\alpha(x_1, x_2), \\ v_3 &= (a_1 x_1 + a_2 x_2 + a_3) x_3 + w_3(x_1, x_2), \end{aligned} \tag{1.18}$$

except for an additive rigid displacement field. Here $\underline{w}$ is an arbitrary vector field independent of x_3, and we have used the notations $a_\alpha = \varepsilon_{\rho\alpha}\beta_\rho$, $a_3 = \alpha_3$, $a_4 = \beta_3$. Let us prove that the functions w_i and the constants $a_s (s=1,2,3,4)$ can be determined so that $\underline{v} \in K_I(F_3, M_1, M_2, M_3)$. The stress-displacement relations imply that

$$\begin{aligned} S_{\alpha\beta}(\underline{v}) &= \lambda(a_\rho x_\rho + a_3)\delta_{\alpha\beta} + T_{\alpha\beta}(\underline{w}), \\ S_{3\alpha}(\underline{v}) &= \mu(w_{3,\alpha} - a_4 \varepsilon_{\alpha\rho} x_\rho),\ S_{33}(\underline{v}) = (\lambda + 2\mu)(a_\rho x_\rho + a_3) + \lambda w_{\rho,\rho}\ . \end{aligned} \tag{1.19}$$

where

$$T_{\alpha\beta}(\underline{w}) = \mu(w_{\alpha,\beta} + w_{\beta,\alpha}) + \lambda\delta_{\alpha\beta}w_{\rho,\rho}. \tag{1.20}$$

The equations of equilibrium and the condition on the lateral boundary reduce to

$$(T_{\alpha\beta}(\underline{w}))_{,\beta} + f_\alpha = 0 \quad \text{on } \Sigma, \quad T_{\alpha\beta}(\underline{w})n_\beta = p_\alpha \text{ on } \Gamma, \tag{1.21}$$

$$\Delta w_3 = 0 \quad \text{on } \Sigma, \quad \frac{\partial w_3}{\partial n} = a_4\varepsilon_{\alpha\beta}n_\alpha x_\beta \text{ on } \Gamma, \tag{1.22}$$

where

$$f_\alpha = \lambda a_\alpha, \quad p_\alpha = -\lambda(a_\rho x_\rho + a_3)n_\alpha. \tag{1.23}$$

The relations (1.20),(1.21) and (1.23) constitute a two-dimensional boundary-value problem (cf.[47], Sect. 45). The necessary and sufficient conditions to solve this problem are

$$\int_\Sigma f_\alpha da + \int_\Gamma p_\alpha ds = 0, \quad \int_\Sigma \varepsilon_{\alpha\beta}x_\alpha f_\beta da + \int_\Gamma \varepsilon_{\alpha\beta}x_\alpha p_\beta ds = 0. \tag{1.24}$$

It follows from (1.23) and the divergence theorem that the conditions (1.24) are satisfied. We observe that the boundary-value problem (1.21) is satisfied if one chooses

$$T_{\alpha\beta}(\underline{w}) = -\lambda(a_\rho x_\rho + a_3)\delta_{\alpha\beta}.$$

Clearly, the above stresses satisfy the compatibility condition. It follows from (1.20) that

$$w_{1,1} = w_{2,2} = -\frac{\lambda}{2(\lambda+\mu)}(a_\rho x_\rho + a_3), \quad w_{1,2} + w_{2,1} = 0.$$

The integration of these equations yields

$$w_\alpha = a_1 w_\alpha^{(1)} + a_2 w_\alpha^{(2)} + a_3 w_\alpha^{(3)},$$

where

$$w_\alpha^{(\beta)} = \nu(\tfrac{1}{2}x_\rho x_\rho \delta_{\alpha\beta} - x_\alpha x_\beta), \quad w_\alpha^{(3)} = -\nu x_\alpha, \tag{1.25}$$

modulo a plane rigid displacement. Here ν designates Poisson's ratio.

It follows from (1.22) that $w_3 = a_4\varphi$, where the function φ is characterized by

$$\Delta\varphi = 0 \quad \text{on } \Sigma, \quad \frac{\partial\varphi}{\partial n} = \varepsilon_{\alpha\beta}n_\alpha x_\beta \text{ on } \Gamma. \tag{1.26}$$

Clearly, the vector field $\underline{v}$ can be written in the form

$$\underline{v} = \sum_{j=1}^{4} a_j \underline{v}^{(j)} \quad , \tag{1.27}$$

where the vectors $\underline{v}^{(j)}$ $(j=1,2,3,4)$ are defined by

$$\begin{aligned} &v_\alpha^{(\beta)} = -\tfrac{1}{2} x_3^2 \delta_{\alpha\beta} + w_\alpha^{(\beta)} , \quad v_3^{(\beta)} = x_\beta x_3 \quad (\beta = 1,2), \\ &v_\alpha^{(3)} = w_\alpha^{(3)} , \quad v_3^{(3)} = x_3 , \quad v_\alpha^{(4)} = \varepsilon_{\beta\alpha} x_\beta x_3 , \quad v_3^{(4)} = \varphi , \end{aligned} \tag{1.28}$$

We note that $\underline{v}^{(j)} \in \mathcal{D}$ $(j=1,2,3,4)$. The conditions on the terminal cross-section Σ_1 furnish the following system for the unknown constants

$$\begin{aligned} E(I_{\alpha\beta} a_\beta + A x_\alpha^0 a_3) &= \varepsilon_{\alpha\beta} M_\beta , \\ EA(a_1 x_1^0 + a_2 x_2^0 + a_3) &= -F_3 , \\ \mu D a_4 &= -M_3 , \end{aligned} \tag{1.29}$$

where A is the area of the cross-section, x_α^0 are the coordinates of the centroid of Σ_1, E designates Young's modulus, and

$$I_{\alpha\beta} = \int_\Sigma x_\alpha x_\beta \, da , \qquad D = \int_\Sigma (x_\alpha x_\alpha + \varepsilon_{\alpha\beta} x_\alpha \varphi_{,\beta}) da . \tag{1.30}$$

If the rectangular Cartesian coordinate frame is chosen in such a way that the x_α-axes are principal centroidal axes of the cross-section Σ_1, then (1.27) and (1.29) lead to the well-known Saint-Venant solution. □

We present Saint-Venant's solutions, which are needed subsequently.

i) Saint-Venant's extension solution:

$$\begin{aligned} &\underline{v} = a_3 \underline{v}^{(3)} , \quad v_\alpha^{(3)} = -\nu x_\alpha , \quad v_3^{(3)} = x_3 , \\ &S_{\alpha\beta}(\underline{v}) = 0 , \quad S_{3\alpha}(\underline{v}) = 0 , \quad S_{33}(\underline{v}) = E a_3 , \end{aligned} \tag{1.31}$$

where

$$F_3 = -EA \, a_3 \quad . \tag{1.32}$$

The relation (1.32) is known as Saint-Venant's formula for extension.

ii) Saint-Venant's bending solution:

$$\underline{v} = a_1\underline{v}^{(1)}, \quad v_1^{(1)} = \frac{1}{2}(\nu x_2^2 - \nu x_1^2 - x_3^2),$$
$$v_2^{(1)} = -\nu x_1 x_2, \quad v_3^{(1)} = x_1 x_3, \tag{1.33}$$
$$S_{\alpha\beta}(\underline{v}) = 0, \quad S_{3\alpha}(\underline{v}) = 0, \quad S_{33}(\underline{v}) = Ea_1 x_1,$$

where

$$M_2 = EI_{11}a_1. \tag{1.34}$$

The relation (1.34) is called Saint-Venant's formula for bending.

iii) Saint-Venant's torsion solution:

$$\underline{v} = a_4\underline{v}^{(4)}, \quad v_\alpha^{(4)} = \varepsilon_{\beta\alpha}x_\beta x_3, \quad v_3^{(4)} = \varphi,$$
$$S_{\alpha\beta}(\underline{v}) = 0, \quad S_{33}(\underline{v}) = 0, \quad S_{3\alpha}(\underline{v}) = \mu a_4(\varphi_{,\alpha} - \varepsilon_{\alpha\rho}x_\rho), \tag{1.35}$$

where

$$M_3 = -\mu D a_4. \tag{1.36}$$

The constant a_4 is known as specific angle of twist, and μD is called the torsional rigidity.
The relation (1.36) is known as Saint-Venant's formula for torsion.

Let us note that the vectors $\underline{v}^{(j)}$ $(j=1,2,3,4)$ defined by (1.27) depend only on the cross-section and the elasticity field. Let $\hat{a}$ be the four-dimensional vector (a_1, a_2, a_3, a_4). We will write $\underline{v}\{\hat{a}\}$ for the displacement vector $\underline{v}$ defined by (1.27), indicating thus its dependence on the constants a_s $(s=1,2,3,4)$.

In view of Corollaries 1.1, 1.2 and Theorem 1.2 it is natural to seek a solution of the problem (P_2) in the form

$$\underline{u}^o = \int_0^{x_3} \underline{v}\{\hat{b}\}dx_3 + \underline{v}\{\hat{c}\} + \underline{w}^o, \tag{1.37}$$

where $\hat{b} = (b_1, b_2, b_3, b_4)$ and $\hat{c} = (c_1, c_2, c_3, c_4)$ are two constant four-dimensional vectors, and $\underline{w}^o$ is a vector field independent of x_3 such that $\underline{w}^o \in C^1(\bar{\Sigma}) \cap C^2(\Sigma)$.

Theorem 1.3. The vector field $\underline{u}^o$ defined by (1.37) is a solution of the problem (P_2) if and only if $\underline{u}^0$ is Saint-Venant's solution.

Proof. Let us prove that the vector field $\underline{w}^0$ and the constants b_s, c_s $(s=1,2,3,4)$ can be determined so that $\underline{u}^0 \in K_{II}(F_1,F_2)$. It is interesting to note that the determination of $\hat{b}$ from the condition $\underline{u}^0 \in K_{II}(F_1,F_2)$ can be made in a simple way. Thus, if $\underline{u}^0 \in K_{II}(F_1,F_2)$, then by Corollary 1.2 and (1.37),

$$\underline{v}\{\hat{b}\} \in K_I(0,F_2,-F_1,0). \tag{1.38}$$

In view of (1.29) and (1.38) we arrive at

$$E(I_{\alpha\beta}b_\beta + A\, x_\alpha^0 b_3) = -F_\alpha\,, \quad b_\rho x_\rho^0 + b_3 = 0\,, \quad b_4 = 0\,. \tag{1.39}$$

It follows from (1.27),(1.28),(1.37) and (1.39) that

$$u_\alpha^0 = -\tfrac{1}{6} b_\alpha x_3^3 - \tfrac{1}{2} c_\alpha x_3^2 - c_4 \varepsilon_{\alpha\beta} x_\beta x_3 + \sum_{j=1}^{3} (c_j + x_3 b_j) w_\alpha^{(j)} + w_\alpha^0\,,$$

$$u_3^0 = \tfrac{1}{2}(b_\rho x_\rho + b_3)x_3^2 + (c_\rho x_\rho + c_3)x_3 + c_4\varphi + \psi\,,$$

where we have used the notation $w_3^0 = \psi$. The stress-displacement relations imply that

$$S_{\alpha\beta}(\underline{u}^0) = T_{\alpha\beta}(\underline{w}^0),$$

$$S_{\alpha 3}(\underline{u}^0) = \mu\left[c_4(\varphi_{,\alpha} - \varepsilon_{\alpha\beta}x_\beta) - \nu x_\alpha(b_\rho x_\rho + b_3) + \tfrac{1}{2} b_\alpha \nu x_\rho x_\rho + \psi_{,\alpha}\right],$$

$$S_{33}(\underline{u}^0) = E\left[(b_\rho x_\rho + b_3)x_3 + c_\rho x_\rho + c_3\right] + \lambda w_{\rho,\rho}^0\,,$$

where

$$T_{\alpha\beta}(\underline{w}^0) = \mu(w_{\alpha,\beta}^0 + w_{\beta,\alpha}^0) + \lambda\delta_{\alpha\beta} w_{\rho,\rho}^0\,. \tag{1.40}$$

The equations of equilibrium and the condition on the lateral boundary reduce to

$$(T_{\alpha\beta}(\underline{w}^0))_{,\beta} = 0 \text{ on } \Sigma\,, \quad T_{\alpha\beta}(\underline{w}^0)\, n_\beta = 0 \text{ on } \Gamma\,, \tag{1.41}$$

$$\Delta\psi = -2(b_\rho x_\rho + b_3) \text{ on } \Sigma\,, \quad \frac{\partial\psi}{\partial n} = b_\alpha \nu x_\rho (x_\alpha n_\rho - \tfrac{1}{2} n_\alpha x_\rho) + b_3 \nu x_\alpha n_\alpha \text{ on } \Gamma\,. \tag{1.42}$$

It follows from (1.40) and (1.41) that $[w_\alpha^0, T_{\alpha\beta}(\underline{w}^0)]$ is a plane elastic state (cf. [47], Sect. 46) corresponding to zero body forces and null boundary data. We conclude that $w_\alpha^0 = 0$ (modulo a plane rigid displacement). Thus, the equations of equilibrium and the condition on the lateral boundary are satisfied if and only if the function ψ is characterized by (1.42) and $w_\alpha^0 = 0$. The necessary and sufficient condition to solve the boundary-value problem (1.42) is satisfied on the basis of the second of (1.39).

The conditions $R_3(\underline{u}^0) = 0$, $\underline{H}(\underline{u}^0) = \underline{0}$ are satisfied if and only if

$$\begin{aligned} &I_{\alpha\beta}c_\beta + A\, x_\alpha^0 c_3 = 0 \quad , \quad c_\rho x_\rho^0 + c_3 = 0 \,, \\ &Dc_4 = -\int_\Sigma \varepsilon_{\alpha\beta}x_\alpha(\psi_{,\beta} + \tfrac{1}{2}\, b_{\beta\nu}x_\rho x_\rho)da \,. \end{aligned} \tag{1.43}$$

Since $H_\alpha(\underline{u}^0_{,3}) = \varepsilon_{\alpha\beta}R_\beta(\underline{u}^0)$ and $\underline{u}^0_{,3} = \underline{v}\{\hat{b}\} \in K_I(0,F_2,-F_1,0)$ it follows that $R_\alpha(\underline{u}^0) = F_\alpha$. We conclude that $\hat{b}$ is determined by (1.39), ψ is characterized by (1.42), $c_i = 0$ and c_4 is given by (1.43). If the rectangular Cartesian coordinate frame is chosen in such a way that x_α-axes are principal centroidal axes of the cross-section Σ_1, then $\underline{u}^0$ reduces to Saint-Venant's solution. □

<u>Remark.</u> We derived the equations (1.37) from the conditions $R_3(\underline{u}^0)=0$, $\underline{H}(\underline{u}^0)=\underline{0}$. If we replace these conditions by $R_3(\underline{u}^0) = F_3$, $\underline{H}(\underline{u}^0) = \underline{M}$, then we arrive at

$$\begin{aligned} &E(I_{\alpha\beta}c_\beta + A\, x_\alpha^0 c_3) = \varepsilon_{\alpha\beta}M_\beta, \quad AE(c_\rho x_\rho + c_3) = -F_3 \,, \\ &\mu Dc_4 = -M_3 - \mu\int_\Sigma \varepsilon_{\alpha\beta}x_\alpha(\psi_{,\beta} + \tfrac{1}{2}\, b_{\beta\nu}x_\rho x_\rho)da \,. \end{aligned} \tag{1.44}$$

Clearly, if $\hat{b}$ is given by (1.39), ψ is characterized by (1.42), and $\hat{c}$ is determined by (1.44), then $\underline{u}^0 \in K(\underline{F},\underline{M})$. Thus, we have the following result:

<u>Theorem 1.4.</u> The vector field $\underline{u}^0$ defined by (1.37) is a solution of the problem (P) if and only if $\underline{u}^0$ is Saint-Venant's solution.

<u>Theorem 1.5.</u> Let $\underline{u}$ be a solution of the problem (P_2). Then $\underline{u}$ admits the decomposition

$$\underline{u} = \underline{u}' + \underline{u}^0 \,, \tag{1.45}$$

where $\underline{u}' \in \mathcal{D}$, $\underline{u}'_{,3} \in K_I(0,F_2,-F_1,0)$ and

$$\underline{u}^o \in K_I(-R_3(\underline{u}'),\ -H_1(\underline{u}'),\ -H_2(\underline{u}'),\ -H_3(\underline{u}')).$$

Proof. Let $\underline{u}' \in \mathfrak{D}$, $\underline{u}'_{,3} \in K_I(0,F_2,-F_1,0)$. In view of Theorem 1.1 we find

$$R_\alpha(\underline{u}') = \varepsilon_{\beta\alpha} H_\beta(\underline{u}'_{,3}) = F_\alpha .$$

Let $\underline{u} \in K_{II}(F_1,F_2)$. If we define $\underline{u}^o$ by $\underline{u}^o = \underline{u} - \underline{u}'$, then $\underline{u}^o \in \mathfrak{D}$ and

$$R_\alpha(\underline{u}^o) = R_\alpha(\underline{u}) - R_\alpha(\underline{u}') = 0 ,$$

$$R_3(\underline{u}^o) = -R_3(\underline{u}') , \ \underline{H}(\underline{u}^o) = -\underline{H}(\underline{u}').$$

Thus, we conclude that the decomposition (1.45) holds. □

We assume for the remainder of this chapter that the x_α-axes are principal centroidal axes of Σ_1. In this case, from $\underline{u}^o \in K_{II}(F,0)$ and (1.39) it follows that $b_1=b$, $b_2=b_3=b_4=0$ where b is given by

$$F = - EI_{11}\ b . \tag{1.46}$$

This is Saint-Venant's formula for flexure.

The above method of deriving Saint-Venant's solutions has been established in [79].

1.4. Minimum Energy Characterizations of Saint-Venant's Solutions

As mentioned before, E.Sternberg and J.K.Knowles [143] characterized Saint-Venant's solutions in terms of certain associated minimum strain-energy properties. Thus, the extension and bending solutions are uniquely determined by the fact that they render the total strain energy an absolute minimum over that subset of the solutions to the respective relaxed problem which results from holding the resultant load or bending couple fixed and from requiring the shearing tractions to vanish pointwise on the ends of the cylinder. Similarly, among all solutions of the relaxed torsion problem that correspond to a fixed torque and to vanishing normal tractions on the ends of the cylinder, Saint-Venant's solution is uniquely distinguished by the fact that it furnishes the absolute minimum of the total strain energy. Sternberg and Knowles have pointed out that the analogous minimum strain-energy characterization of Saint-Venant's flexure solution holds true if and only if Poisson's ratio happens to be zero. Other results concerning the status of Saint-Venant's solutions as minimizers of energy have been established by O. Maisonneuve [99] and J.L. Ericksen [33]. In this section, we present the result of E. Sternberg and J.K. Knowles [143] concerning the minimum strain-energy characterizations of Saint-Venant's extension, bending and torsion solutions. An analogous minimum strain-energy characterization [79] of the partial derivative with respect to the axial coordinate of Saint-Venant's flexure solution is also established.

Let Y_E denote the set of all equilibrium displacement fields $\underline{u}$ that satisfy the conditions

$$\underline{s}(\underline{u}) = \underline{0} \text{ on } \Pi, \quad S_{3\beta}(\underline{u}) = 0 \text{ on } \Sigma_\alpha, \quad R_3(\underline{u}) = F. \tag{1.47}$$

<u>Theorem 1.6.</u> Let $\underline{v}$ be Saint-Venant's extension solution corresponding to a scalar load F. Then

$$U(\underline{v}) \leq U(\underline{u}),$$

for every $\underline{u} \in Y_E$, and equality holds only if $\underline{u} = \underline{v}$ modulo a rigid displacement.

<u>Proof.</u> Let $\underline{u} \in Y_E$ and define

$$\underline{u}' = \underline{u} - \underline{v}. \tag{1.48}$$

Then $\underline{u}'$ is an equilibrium displacement field that satisfies

$$\underline{s}(\underline{u}') = \underline{0} \text{ on } \Pi, \quad S_{3\beta}(\underline{u}') = 0 \text{ on } \Sigma_\alpha, \quad R_3(\underline{u}') = 0. \tag{1.49}$$

By (1.5),(1.6) and (1.48),

$$U(\underline{u}) = U(\underline{u}') + U(\underline{v}) + \langle \underline{u}', \underline{v} \rangle .$$

It follows from (1.8),(1.9),(1.14),(1.31) and (1.49) that

$$\langle \underline{u}', \underline{v} \rangle = \int_{\Sigma_2} S_{3i}(\underline{u}')v_i da - \int_{\Sigma_1} S_{3i}(\underline{u}')v_i da = -a_3\, h\, R_3(\underline{u}') = 0 .$$

Thus $U(\underline{u}) \geq U(\underline{v})$, and $U(\underline{u}) = U(\underline{v})$ only if $\underline{u}'$ is a rigid displacement.□

Let Y_B denote the set of all equilibrium displacement fields $\underline{u}$ that satisfy the conditions

$$\underline{s}(\underline{u}) = \underline{0} \text{ on } \Pi, \quad S_{3\beta}(\underline{u}) = 0 \text{ on } \Sigma_\alpha, \quad H_2(\underline{u}) = M_2 . \tag{1.50}$$

Theorem 1.7. Let $\underline{v}$ be Saint-Venant's bending solution corresponding to a couple of scalar moment M_2. Then

$$U(\underline{v}) \leq U(\underline{u}),$$

for every $\underline{u} \in Y_B$, and equality holds only if $\underline{u} = \underline{v}$ modulo a rigid displacement.

Proof. Let $\underline{u} \in Y_B$. Since $\underline{v} \in Y_B$ it follows that the field

$$\underline{u}' = \underline{u} - \underline{v} ,$$

is an equilibrium displacement field that satisfies

$$\underline{s}(\underline{u}') = \underline{0} \text{ on } \Pi, \quad S_{3\beta}(\underline{u}') = 0 \text{ on } \Sigma_\alpha, \quad H_2(\underline{u}') = 0 . \tag{1.51}$$

According to (1.8),(1.9),(1.14),(1.33) and (1.51) we have

$$\langle \underline{u}', \underline{v} \rangle = \int_{\Sigma_2} S_{33}(\underline{u}') v_3 da - \int_{\Sigma_1} S_{33}(\underline{u}') v_3 da = h\, a_1\, H_2(\underline{u}') = 0 .$$

Thus,

$$U(\underline{u}) = U(\underline{u}') + U(\underline{v}).$$

The desired conclusion is now immediate. □

Remark. It is a simple matter to verify that the above minimum strain-energy characterizations also hold if the conditions

$$S_{3\beta}(\underline{u}) = 0 \quad \text{on} \quad \Sigma_\alpha ,$$

which appear in (1.47) and (1.50) are replaced by

$$R_\alpha(\underline{u}) = 0, \; [S_{3\beta}(\underline{u})](x_1,x_2,h) = [S_{3\beta}(\underline{u})](x_1,x_2,0), \; (x_1,x_2) \in \Sigma_1 .$$

Let Y_T denote the set of all equilibrium displacement fields $\underline{u}$ that satisfy the conditions

$$\underline{s}(\underline{u}) = \underline{0} \text{ on } \Pi , \quad S_{33}(\underline{u}) = 0 \text{ on } \Sigma_\alpha , \; H_3(\underline{u}) = M_3 . \tag{1.52}$$

Theorem 1.8. Let $\underline{v}$ be Saint-Venant's torsion solution corresponding to the scalar torque M_3. Then

$$U(\underline{v}) \leq U(\underline{u}),$$

for every $\underline{u} \in Y_T$, and equality holds only if $\underline{u} = \underline{v}$ modulo a rigid displacement.

Proof. Note that $\underline{v} \in Y_T$. Let $\underline{u} \in Y_T$, and define $\underline{u}'$ by $\underline{u}' = \underline{u} - \underline{v}$. Then $\underline{u}'$ is an equilibrium displacement field such that

$$\underline{s}(\underline{u}') = \underline{0} \text{ on } \Pi, \quad S_{33}(\underline{u}') = 0 \text{ on } \Sigma_\alpha, \quad H_3(\underline{u}') = 0. \tag{1.53}$$

If we apply (1.8) and (1.9), we conclude, with the aid of (1.35) and (1.53), that

$$\langle \underline{u}', \underline{v} \rangle = -a_4 h \int_{\Sigma_2} \varepsilon_{\alpha\beta} x_\beta S_{3\alpha}(\underline{u}')da = -a_4 h\, H_3(\underline{u}') = 0 .$$

This result implies that

$$U(\underline{u} - \underline{v}) = U(\underline{u}) - U(\underline{v}). \tag{1.54}$$

The proof follows from (1.54). □

Remark. If we replace in (1.52) the conditions

$$S_{33}(\underline{u}) = 0 \quad \text{on } \Sigma_\alpha \quad ,$$

by

$$[S_{33}(\underline{u})](x_1,x_2,h) = [S_{33}(\underline{u})](x_1,x_2,0), \quad (x_1,x_2) \in \Sigma \quad ,$$

the above theorem also remains valid.

Let Y_F denote the set of all equilibrium displacement fields $\underline{u}$ that satisfy the conditions

$$\begin{gathered} \underline{u}_{,3} \in C^1(\bar{B}), \qquad \underline{s}(\underline{u}) = \underline{0} \text{ on } \Pi \ , \ R_\alpha(\underline{u}) = F_\alpha \ , \\ [S_{3\beta}(\underline{u}_{,3})](x_1,x_2,h) = [S_{3\beta}(\underline{u}_{,3})](x_1,x_2,0), (x_1,x_2) \in \Sigma \quad . \end{gathered} \tag{1.55}$$

Theorem 1.9. Let $\underline{u}^0$ be Saint-Venant's flexure solution corresponding to the scalar loads F_1 and F_2. Then

$$U(\underline{u}^o_{,3}) \leq U(\underline{u}_{,3}),$$

for every $\underline{u} \in Y_F$, and equality holds only if $\underline{u}_{,3} = \underline{u}^o_{,3}$ (modulo a rigid displacement).

Proof. Let $\underline{u} \in Y_F$ and define $\underline{u}' = \underline{u} - \underline{u}^o$. Then $\underline{u}'$ is an equilibrium displacement field that satisfies

$$\underline{u}'_{,3} \in C^1(\bar{B}), \qquad \underline{s}(\underline{u}') = \underline{0} \text{ on } \Pi, \quad R_\alpha(\underline{u}') = 0,$$
$$[S_{3\beta}(\underline{u}'_{,3})](x_1,x_2,h) = [S_{3\beta}(\underline{u}'_{,3})](x_1,x_2,0), \quad (x_1,x_2) \in \Sigma \quad . \tag{1.56}$$

According to (1.5),(1.37) and Theorem 1.2, we have

$$U(\underline{u}_{,3}) = U(\underline{u}'_{,3} + \underline{u}^o_{,3}) = U(\underline{u}'_{,3} + \underline{v}\{\hat{b}\}) = U(\underline{u}'_{,3}) + U(\underline{u}^o_{,3}) + \langle \underline{u}'_{,3}, \underline{v}\{\hat{b}\}\rangle .$$

In view of Theorem 1.1, (1.14) and (1.56), we find that

$$\langle \underline{u}'_{,3}, \underline{v}\{\hat{b}\}\rangle = -\frac{1}{2} b_\alpha h^2 R_\alpha(\underline{u}'_{,3}) + h[b_1 H_2(\underline{u}'_{,3}) - b_2 H_1(\underline{u}'_{,3})] = 0.$$

Thus,

$$U(\underline{u}_{,3}) - U(\underline{u}^o_{,3}) = U(\underline{u}'_{,3} - \underline{v}\{\hat{b}\}).$$

The desired conclusion is immediate. □

The results presented in this section are useful in the study of Truesdell's problem.

Remark. The above results concerning the minimum strain-energy characterizations of Saint-Venant's solutions are based on a comparison with a subset-rather than with the complete class of solutions to the corresponding relaxed problem. It is natural to seek also those members of the class of solutions to each of the four relaxed problems that minimize the strain energy over the complete class of solutions to the corresponding relaxed problem.

1.5. Truesdell's Problem

In Saint-Venant's solution of the torsion problem, corresponding to a couple of scalar moment M_3, the specific angle of twist a_4 is given by (1.36). Let K_T denote the set of all displacement fields that correspond to the solutions of the foregoing torsion problem. In [148 - 150], C.Truesdell proposed the following problem: to define the functional $\tau(\cdot)$ on K_T such that

$$M_3 = -\mu D\tau(\underline{u}) \ , \quad \text{for each } \underline{u}\in K_T \ .$$

Following W.A.Day [25], $\tau(\underline{u})$ is called the generalized twist at $\underline{u}$. In [25], W.A.Day established a solution of Truesdell's problem. P.Podio-Guidugli [123] solved Truesdell's problem rephrased for extension and bending. In [79], we have studied Truesdell's problem rephrased for flexure. In this section we present these results.

With a view toward a concise presentation of Day's solution we denote by Q_T the set of all equilibrium displacement fields $\underline{u}$ that satisfy the conditions

$$\begin{aligned} &\underline{s}(\underline{u}) = \underline{0} \ \text{ on } \Pi \ , \quad && S_{33}(\underline{u}) = 0 \ \text{ on } \Sigma_\alpha \ , \\ &R_\alpha(\underline{u}) = 0, && H_3(\underline{u}) = M_3 \ . \end{aligned} \tag{1.57}$$

If $\underline{u}\in Q_T$, then $R_3(\underline{u}) = 0$, $H_\alpha(\underline{u}) = 0$ so that $\underline{u}\in K_T$. W.A.Day considered the real function

$$\alpha \longrightarrow \|\underline{u} - \alpha\underline{v}^{(4)}\|_e^2 \ , \tag{1.58}$$

where $\underline{u}\in Q_T$ and $\underline{v}^{(4)}$ is the displacement field given by (1.35). The field $\alpha\underline{v}^{(4)}$ is called the torsion field with twist α.

Clearly, the function defined in (1.58) attains its minimum at

$$\gamma(\underline{u}) = \frac{\langle \underline{u},\underline{v}^{(4)}\rangle}{\|\underline{v}^{(4)}\|_e^2} \ . \tag{1.59}$$

Thus, $\gamma(\underline{u})$ is the twist of that torsion field which approximates $\underline{u}$ most closely. Let us prove that

$$\gamma(\underline{u}) = \tau(\underline{u}), \ \text{for every } \underline{u}\in Q_T \ .$$

In view of (1.8),(1.9),(1.14),(1.35) and (1.57) we arrive at

$$\langle \underline{u},\underline{v}^{(4)}\rangle = \int_{\partial B}\underline{s}(\underline{u})\cdot\underline{v}^{(4)}da = \int_{\Sigma_2}[h\varepsilon_{\beta\alpha}x_\beta S_{3\alpha}(\underline{u}) + \varphi S_{33}(\underline{u})]da =$$

$$= h\int_{\Sigma_2} \varepsilon_{\beta\alpha} x_\beta S_{3\alpha}(\underline{u})da = -h\, H_3(\underline{u}),$$

$$\| \underline{v}^{(4)} \|_e^2 = h\int_{\Sigma_2} \varepsilon_{\beta\alpha} x_\beta S_{3\alpha}(\underline{v}^{(4)})da = \mu h D\,. \tag{1.60}$$

It follows from (1.59) and (1.60) that

$$H_3(\underline{u}) = -\mu D \gamma(\underline{u}),$$

for any $\underline{u} \in Q_T$. Clearly, $\gamma(\underline{u}) = \tau(\underline{u})$ for each $\underline{u} \in Q_T$. Thus, Saint-Venant's formula (1.36) applies to the displacement fields $\underline{u}$ which belong to Q_T.

On the other hand, by (1.8) and (1.35) we find that

$$<\underline{u},\underline{v}^{(4)}> = \mu\Big[\int_{\Sigma_2} u_\alpha(\varphi_{,\alpha} - \varepsilon_{\alpha\rho}x_\rho)da - \int_{\Sigma_1} u_\alpha(\varphi_{,\alpha} - \varepsilon_{\alpha\rho}x_\rho)da\Big]. \tag{1.61}$$

We conclude from (1.59),(1.60) and (1.61) that the generalized twist $\tau(\underline{u})$ associated with any $\underline{u} \in Q_T$ is given by

$$\tau(\underline{u}) = \frac{1}{hD}\Big[\int_{\Sigma_2} u_\alpha(\varphi_{,\alpha} - \varepsilon_{\alpha\rho}x_\rho)da - \int_{\Sigma_1} u_\alpha(\varphi_{,\alpha} - \varepsilon_{\alpha\rho}x_\rho)da\Big].$$

<u>Remark.</u> Since div $\underline{v}^{(4)} = 0$, it follows that

$$<\underline{u},\underline{v}^{(4)}> = \mu\int_B \underline{E}(\underline{u})\cdot\underline{E}(\underline{v}^{(4)})dv\,.$$

Thus, the energy norm which appears in (1.58) can be replaced by the strain norm. In [25], Day used the strain norm.

Let us consider now Saint-Venant's formula (1.32). Truesdell's problem can be set also for the extension problem. Let Q_E denote the set of all equilibrium displacement fields $\underline{u}$ that satisfy the conditions

$$\underline{s}(\underline{u}) = \underline{0} \ \text{ on } \Pi, \quad S_{3\beta}(\underline{u}) = 0 \ \text{ on } \Sigma_\alpha,$$

$$H_\alpha(\underline{u}) = 0\,, \qquad R_3(\underline{u}) = F_3\,. \tag{1.62}$$

Clearly, if $\underline{u} \in Q_E$, then $R_\alpha(\underline{u}) = 0$, $H_3(\underline{u}) = 0$, so that

$\underline{u} \in K_I(F_3,0,0,0)$. P.Podio-Guidugli [123] considered the function

$$\beta \longrightarrow \| \underline{u} - \beta \underline{v}^{(3)} \|_e^2 , \tag{1.63}$$

where $\underline{u} \in Q_E$ and $\underline{v}^{(3)}$ is the displacement field given by (1.31). The field $\beta \underline{v}^{(3)}$ is called the extension field with axial strain β. The function defined in (1.63) attains its minimum at

$$\mathcal{E}(\underline{u}) = \frac{< \underline{u}, \underline{v}^{(3)} >}{\| \underline{v}^{(3)} \|_e^2} . \tag{1.64}$$

It follows from (1.8),(1.9),(1.14),(1.31) and (1.62) that

$$< \underline{u}, \underline{v}^{(3)} > = \int_{\partial B} \underline{s}(\underline{u}) \cdot \underline{v}^{(3)} da = h \int_{\Sigma_2} S_{33}(\underline{u}) da = -h\, R_3(\underline{u}).$$
$$\| \underline{v}^{(3)} \|_e^2 = h\, EA . \tag{1.65}$$

By (1.64) and (1.65),

$$R_3(\underline{u}) = -EA\,\mathcal{E}(\underline{u}),$$

for each $\underline{u} \in Q_E$. Thus, Saint-Venant's formula (1.32) applies to any displacement field $\underline{u} \in Q_E$. We call $\mathcal{E}(\underline{u})$ the generalized axial strain associated with the displacement field $\underline{u}$. In view of (1.8) and (1.31) we have

$$< \underline{u}, \underline{v}^{(3)} > = \int_{\partial B} \underline{s}(\underline{v}^{(3)}) \cdot \underline{u}\, da = E(\int_{\Sigma_2} u_3\, da - \int_{\Sigma_1} u_3\, da). \tag{1.66}$$

Thus, by (1.64),(1.65) and (1.66) we conclude that the generalized axial strain $\mathcal{E}(\underline{u})$ associated with any $\underline{u} \in Q_E$ is given by

$$\mathcal{E}(\underline{u}) = \frac{1}{hA} (\int_{\Sigma_2} u_3\, da - \int_{\Sigma_1} u_3\, da).$$

We now consider the bending problem. Let Q_B denote the set of all equilibrium displacement fields $\underline{u}$ that satisfy the conditions

$$\underline{s}(\underline{u}) = \underline{0} \quad \text{on } \Pi , \qquad S_{3\beta}(\underline{u}) = 0 \quad \text{on } \Sigma_\alpha ,$$
$$R_3(\underline{u}) = 0 , \quad H_1(\underline{u}) = 0 , \quad H_2(\underline{u}) = M_2 .$$

If $\underline{u} \in Q_B$, then $\underline{u} \in K_I(0,0,M_2,0)$. In the same manner, we are led to the generalized axial curvature $\varkappa(\underline{u})$, associated with any $\underline{u} \in Q_B$,

$$\varkappa(\underline{u}) = \frac{1}{hI_{11}} \left(\int_{\Sigma_2} x_1 u_3 da - \int_{\Sigma_1} x_1 u_3 da \right).$$

Moreover, the formula of Saint-Venant type

$$H_2(u) = E\, I_{11} \varkappa(\underline{u}),$$

applies for each $\underline{u} \in Q_B$.

Let us study Truesdell's problem for flexure. Let Q_F denote the set of all equilibrium displacement fields $\underline{u}$ that satisfy the conditions

$$\begin{gathered} \underline{u}_{,3} \in C^1(\bar{B}), \quad S_{3\beta}(\underline{u}_{,3}) = 0 \quad \text{on } \Sigma_\alpha, \\ \underline{s}(\underline{u}) = \underline{0} \quad \text{on } \Pi, \quad R_1(\underline{u}) = F, \quad R_2(\underline{u}) = 0, \ R_3(\underline{u}) = 0, \\ \underline{H}(\underline{u}) = \underline{0}. \end{gathered} \tag{1.67}$$

Clearly, if $\underline{u} \in Q_F$ then $\underline{u} \in K_{II}(F,0)$. Moreover, if $\underline{u} \in Q_F$, then by Corollary 1.2, $\underline{u}_{,3} \in K_I(0,0,-F,0)$ and $S_{3\varrho}(\underline{u}_{,3}) = 0$ on Σ_α. In view of Theorem 1.9, we are led to consider the function

$$\xi \longrightarrow \| \underline{u}_{,3} - \xi \underline{v}^{(1)} \|_e^2, \tag{1.68}$$

where $\underline{u} \in Q_F$ and $\underline{v}^{(1)}$ is the displacement field defined by (1.33). Clearly, the function defined in (1.68) attains minimum at

$$\eta(\underline{u}) = \frac{\langle \underline{u}_{,3}, \underline{v}^{(1)} \rangle}{\| \underline{v}^{(1)} \|_e^2}. \tag{1.69}$$

By (1.8),(1.9),(1.14),(1.33) and (1.67) we find

$$\langle \underline{u}_{,3}, \underline{v}^{(1)} \rangle = \int_{\partial B} \underline{s}(\underline{u}_{,3}) \cdot \underline{v}^{(1)} da = h\, H_2(\underline{u}_{,3}) - h^2\, R_1(\underline{u}_{,3}).$$

In view of Theorem 1.1 we arrive at

$$\langle \underline{u}_{,3}, \underline{v}^{(1)} \rangle = -h\, R_1(\underline{u}). \tag{1.70}$$

A simple calculation using $S_{3i}(\underline{v}^{(1)}) = Ex_1 \delta_{i3}$ yields

$$\| \underline{v}^{(1)} \|_e^2 = h\, E\, I_{11} \; . \tag{1.71}$$

We conclude from (1.69),(1.70) and (1.71) that

$$R_1(\underline{u}) = - E\, I_{11} \eta(\underline{u}),$$

for every $\underline{u} \in Q_F$. Thus, we have obtained a formula of Saint-Venant type applicable to any displacement field $\underline{u} \in Q_F$.

On the other hand, by (1.8) we arrive at

$$<\underline{u}_{,3}, \underline{v}^{(1)}> = \int_{\partial B} \underline{s}(\underline{v}^{(1)}) \cdot \underline{u}_{,3} \, da = E\Big(\int_{\Sigma_2} x_1 u_{3,3} da - \int_{\Sigma_1} x_1 u_{3,3} da \Big). \tag{1.72}$$

Thus we conclude from (1.69),(1.71) and (1.72) that

$$\eta(\underline{u}) = \frac{1}{hI_{11}} \Big(\int_{\Sigma_2} x_1 u_{3,3} da - \int_{\Sigma_1} x_1 u_{3,3} da \Big),$$

and interpret the right-hand side as the global measure of strain appropriate to flexure, associated with the displacement field $\underline{u} \in Q_F$. The results presented in this section are related to the results concerning the minimum energy characterizations of corresponding Saint-Venant's solutions.

1.6. Saint-Venant's Principle

The broader significance of Saint-Venant's solutions to the relaxed problem for load distributions that are statically equivalent to, but distinct from those implied by Saint-Venant's results, depends on the validity of the principle bearing his name. Saint-Venant's principle was originally enunciated in order to justify the use of Saint-Venant's solutions. This principle is usually taken to mean that a system of loads having zero resultant force and moment at each end produces a strain field that is negligible away from the ends. The first general statement of Saint-Venant's principle was given by Boussinesq [13] : "An equilibrated system of external forces applied to an elastic body, all of the points of application lying within a given sphere, produces deformations of negligible magnitude at distances from the sphere which are sufficiently large compared to its radius". v.Mises [106] pointed out that this formulation of Saint-Venant's principle is ambiguous, and suggested an amended version of the principle. A rigorous Saint-Venant principle should give sufficient conditions under which the internal effects (stress, strain etc.) will decrease in some specified sense with distance from the load region of the boundary.

The first precise general treatment of any version of Saint-Venant's principle was that of E.Sternberg [142], who formulated and proved the version suggested by von Mises. Two alternative versions of Saint-Venant's principle were established by R.A.Toupin [146] and J.K.Knowles [84]. These authors arrived at estimates for the strain energy U_z contained in that portion of the body which lies beyond a distance z from the load region. The idea of using U_z in the formulation of Saint-Venant's principle is due to Zanaboni [156-158]. Knowles' results are confined to the case of the two-dimensional plane problems. Toupin considered the problem of an anisotropic elastic cylinder of arbitrary length subject to self-equilibrated surface tractions on one of its ends, and free of surface tractions on the remainder of its boundary. In [39], G.Fichera extended Toupin's result to the case of an elastic cylinder subject to self-equilibrated surface tractions on each of its ends, and free of surface traction on the lateral boundary. Clearly, this is the case involved by Saint-Venant's conjecture.

Recently, a number of results have appeared in the literature concerned with a nonlinear version of Saint-Venant's principle. We mention the articles by J.J.Roseman [129], S.Breuer and J.J.Roseman [14], R.G.Muncaster [108], C.O.Horgan and J.K.Knowles [51], R.J.Knops and L.E.Payne [83].

For the history of the problem and the detailed analysis of various results on Saint-Venant's principle we refer to the works of M.E.Gurtin [47], G.Y.Djanelidze [28], G.Fichera [38], C.O.Horgan and J.K.Knowles [53], G.Fichera [40].

In this section we present the results due to Toupin [146] and Fichera [39], which provide the mathematical formulation and proof of Saint-Venant's principle in the context for which it was originally intended.

Let $\underline{u}'$ be Saint-Venant's solution of the relaxed Saint-Venant's problem, and let $\underline{u}''$ be the solution of Saint-Venant's problem. We define the displacement field $\underline{u}$ on B by $\underline{u} = \underline{u}'' - \underline{u}'$. Then, $\underline{u}$ is an equilibrium displacement field that satisfies the conditions

$$\underline{s}(\underline{u}) = \underline{0} \quad \text{on } \Pi ,$$
$$\int_{\Sigma_\alpha} \underline{s}(\underline{u})da = \underline{0} , \quad \int_{\Sigma_\alpha} \underline{x} \times \underline{s}(\underline{u})da = \underline{0} \quad (\alpha = 1,2). \tag{1.73}$$

Thus, $\underline{u}$ is a displacement field corresponding to null body forces and to surface tractions which vanish on the lateral boundary and are self-equilibrated at each end.

Let B_z denote the cylinder defined by

$$B_z = \{\underline{x} : (x_1,x_2)\in\Sigma , \ z < x_3 < h - z\} \ (0 \leq z < \tfrac{h}{2}) . \tag{1.74}$$

We denote by $U_z(\underline{u})$ the strain energy corresponding to the equilibrium displacement field $\underline{u}$ on B_z , i.e.

$$U_z(\underline{u}) = \frac{1}{2}\int_{B_z} \nabla \underline{u}\cdot\underline{C}[\nabla \underline{u}]dv . \tag{1.75}$$

The positive-definiteness of $\underline{C}$ implies that $U_z(\underline{u})$ is a non-increasing function of z.

<u>Theorem 1.10.</u> Assume that B is homogeneous and anisotropic, and assume that the elasticity tensor is symmetric and positive definite. Let $\underline{u}$ be an equilibrium displacement field that satisfies the conditions (1.73). Then the strain energy $U_z(\underline{u})$ satisfies the inequality

$$U_z(\underline{u}) \leq U_0(\underline{u}) \ e^{-(z-\ell)/k(\ell)} \quad (z \geq \ell) , \tag{1.76}$$

for any $\ell > 0$, where

$$k(\ell) = \sqrt{\frac{\mu_M}{\lambda(\ell)}} ,$$

μ_M is the maximum elastic modulus, and $\lambda(\ell)$ is the lowest non-zero characteristic value of free vibration for a slice of the cylinder, of thickness ℓ, taken normal to its generators and that has its boundary traction-free.

<u>Proof.</u> By (1.5),(1.8) and (1.73),

$$U_z(\underline{u}) = \frac{1}{2}\int_{\partial B_z} \underline{s}(\underline{u})\cdot\underline{u} \ da = \frac{1}{2}\Big\{ \int_{S_{h-z}} u_i S_{3i}(\underline{u})da - \int_{S_z} u_i S_{3i}(\underline{u})da\Big\}. \tag{1.77}$$

Here S_z denote the cross-section located at $x_3 = z$.

The resultant force and resultant moment on every part of the cylinder must vanish in equilibrium. Let $B(t_1,t_2)$ $(0 \leq t_1 < t_2 \leq h)$ denote the cylinder

$$B(t_1,t_2) = \{\underline{x} : (x_1,x_2) \in \Sigma ,\ t_1 < x_3 < t_2\}.$$

The conditions of equilibrium for the parts $B(0,z)$ and $B(h-z,h)$ of the cylinder, and the conditions (1.73) imply that

$$\begin{aligned} &\int_{S_z} \underline{s}(\underline{u})da = \underline{0} ,\quad \int_{S_z} \underline{x} \times \underline{s}(\underline{u})da = \underline{0} , \\ &\int_{S_{h-z}} \underline{s}(\underline{u})da = \underline{0} ,\quad \int_{S_{h-z}} \underline{x} \times \underline{s}(\underline{u})da = \underline{0} . \end{aligned} \tag{1.78}$$

Let us introduce the vector fields $\underline{u}^{(\alpha)}$ $(\alpha = 1,2)$ defined by

$$\underline{u}^{(\alpha)} = \underline{u} + \underline{a}^{(\alpha)} + \underline{b}^{(\alpha)} \times \underline{x} \quad (\alpha = 1,2), \tag{1.79}$$

where $\underline{a}^{(\alpha)}$ and $\underline{b}^{(\alpha)}$ are arbitrary constant vectors. Clearly, the vector fields $\underline{u}^{(\alpha)}$ differ from $\underline{u}$ by a rigid displacement. In view of (1.78), the displacement field $\underline{u}$ which appears in the integrands of (1.77) may be replaced by $\underline{u}^{(\alpha)}$ such that

$$U_z(\underline{u}) = \frac{1}{2}\Big\{ \int_{S_{h-z}} u_i^{(1)} s_{3i}(\underline{u})da - \int_{S_z} u_i^{(2)} s_{3i}(\underline{u})da \Big\}. \tag{1.80}$$

If we apply the Schwarz inequality, we find that

$$U_z(\underline{u}) \leq \frac{1}{2}\Big\{\Big(\int_{S_{h-z}} |\underline{u}^{(1)}|^2 da \int_{S_{h-z}} |\underline{S}(\underline{u})|^2 da\Big)^{1/2} + \Big(\int_{S_z} |\underline{u}^{(2)}|^2 da \int_{S_z} |\underline{S}(\underline{u})|^2 da\Big)^{1/2}\Big\}, \tag{1.81}$$

where $|\underline{S}| = \sqrt{\underline{S}\cdot\underline{S}}$.

From the geometric-arithmetic mean inequality, we have

$$\sqrt{ab} \leq \frac{1}{2} \left(\alpha a + \frac{b}{\alpha}\right),$$

for all non-negative scalars a,b and α with $\alpha > 0$. If we apply this inequality to (1.81), we obtain

$$U_z(\underline{u}) \leq \frac{1}{4}\Big\{\alpha \int_{S_{h-z}} |\underline{S}(\underline{u})|^2 da + \frac{1}{\alpha} \int_{S_{h-z}} |\underline{u}^{(1)}|^2 da + \alpha \int_{S_z} |\underline{S}(\underline{u})|^2 da + \frac{1}{\alpha}\int_{S_z} |\underline{u}^{(2)}|^2 da \Big\}, \tag{1.82}$$

where α is an arbitrary positive constant.

Since $\underline{C}$ is symmetric and positive definite, the characteristic values of $\underline{C}$ (considered as a linear transformation on the six-dimensional space of all symmetric tensors) are all strictly positive. Following Toupin [146], we call the largest characteristic value the maximum elastic modulus, the smallest the minimum elastic modulus. We denote the maximum and minimum elastic modulus by μ_M and μ_m, respectively. It follows that

$$\mu_m |\underline{A}|^2 \leq \underline{A}\cdot\underline{C}[\underline{A}] \leq \mu_M |\underline{A}|^2 , \qquad (1.83)$$

for any symmetric tensor $\underline{A}$. It is convenient to introduce the notation

$$W(\underline{u}) = \frac{1}{2}\nabla\underline{u}\cdot\underline{C}[\nabla\underline{u}]. \qquad (1.84)$$

Since the characteristic values of $\underline{C}^2$ are the square of the characteristic values of $\underline{C}$, we have

$$|\underline{S}(\underline{u})|^2 = \underline{C}[\nabla\underline{u}]\cdot\underline{C}[\nabla\underline{u}] = \nabla\underline{u}\cdot\underline{C}^2[\nabla\underline{u}] < \mu_M \nabla\underline{u}\cdot\underline{C}[\nabla\underline{u}] = 2\mu_M W(\underline{u}). \qquad (1.85)$$

By (1.82) and (1.85),

$$U_z(\underline{u}) \leq \frac{1}{4}\Big\{2\alpha\mu_M \int_{S_{h-z}} W(\underline{u})da + \frac{1}{\alpha}\int_{S_{h-z}} |\underline{u}^{(1)}|^2 da + 2\alpha\mu_M \int_{S_z} W(\underline{u})da + \frac{1}{\alpha}\int_{S_z} |\underline{u}^{(2)}|^2 da. \qquad (1.86)$$

Next, choose ℓ such that $0 < \ell < \frac{h}{2}$ and let

$$Q(z,\ell) = \frac{1}{\ell}\int_z^{z+\ell} U_t(\underline{u})dt \quad . \qquad (1.87)$$

If we integrate the inequality (1.86) between the limits z and $z+\ell$, then we find that

$$\ell Q(z,\ell) \leq \frac{1}{4}\Big\{2\alpha\mu_M \int_{V_1} W(\underline{u})dv + \frac{1}{\alpha}\int_{V_1} |\underline{u}^{(1)}|^2 dv + 2\alpha\mu_M \int_{V_2} W(\underline{u})dv + \frac{1}{\alpha}\int_{V_1} |\underline{u}^{(2)}|^2 dv, \qquad (1.88)$$

where

$$V_1 = B(h-z-\ell, h-z), \quad V_2 = B(z, z+\ell).$$

Let $\lambda(\ell)$ denote the lowest non-zero characteristic value of free vibration for the portion $V = B(0,\ell)$ of B. According to the minimum principle from the theory of free vibrations (cf.[47],Sect.76),

$$\lambda(\ell)\int_V \underline{v}^2 dv \leq 2\int_V W(\underline{v})dv \ ;$$

for every $\underline{v} \in C^1(\bar{V}) \cap C^2(V)$ that satisfies

$$\int_V \underline{v}^2 dv \neq 0 \ , \ \int_V \underline{v}\, dv = \underline{0} \ , \ \int_V \underline{x} \times \underline{v}\, dv = \underline{0} \ .$$

The constant vectors $\underline{a}^{(\alpha)}$ and $\underline{b}^{(\alpha)}$ $(\alpha = 1,2)$ can be chosen so that

$$\int_{V_1} \underline{u}^{(1)} dv = \underline{0} \ , \quad \int_{V_1} \underline{x} \times \underline{u}^{(1)} dv = \underline{0} \ ,$$

$$\int_{V_2} \underline{u}^{(2)} dv = \underline{0} \ , \quad \int_{V_2} \underline{x} \times \underline{u}^{(2)} dv = \underline{0} \ .$$

Thus,

$$\begin{aligned} \int_{V_1} |\underline{u}^{(1)}|^2 dv &\leq \frac{2}{\lambda(\ell)}\int_{V_1} W(\underline{u})dv \ , \\ \int_{V_2} |\underline{u}^{(2)}|^2 dv &\leq \frac{2}{\lambda(\ell)}\int_{V_2} W(\underline{u})dv \ . \end{aligned} \tag{1.89}$$

It follows from (1.88) and (1.89) that

$$\ell Q(z,\ell) \leq \frac{1}{2}\left(\alpha\mu_M + \frac{1}{\alpha\lambda(\ell)}\right)\left\{\int_{V_1} W(\underline{u})da + \int_{V_2} W(\underline{u})dv\right\}. \tag{1.90}$$

In view of (1.75),(1.84) and (1.87), we obtain

$$\ell\frac{\partial}{\partial z} Q(z,\ell) = U_{z+\ell}(\underline{u}) - U_z(\underline{u}) = -\int_{V_1} W(\underline{u})dv - \int_{V_2} W(\underline{u})dv \ . \tag{1.91}$$

By (1.90) and (1.91),

$$g(\alpha,\ell)\frac{\partial}{\partial z} Q(z,\ell) + Q(z,\ell) \leq 0 \ , \tag{1.92}$$

where

$$g(\alpha,\ell) = \frac{1}{2}\left(\alpha\mu_M + \frac{1}{\alpha\lambda(\ell)}\right) .$$

From the geometric-arithmetic mean inequality, we have

$$g(\alpha,\ell) \geq \sqrt{\frac{\mu_M}{\lambda(\ell)}} = k(\ell),$$

for any $\alpha > 0$. Thus, (1.92) implies that

$$k(\ell)\,\frac{\partial}{\partial z}\,Q(z,\ell) + Q(z,\ell) \leq 0.$$

Therefore, one has

$$\frac{\partial}{\partial z}\,(e^{z/k(\ell)}\,Q(z,\ell)) \leq 0\ . \tag{1.93}$$

It follows from (1.93) that

$$Q(t_2,\ell) \leq e^{-(t_2-t_1)/k(\ell)}\,Q(t_1,\ell)\ , \tag{1.94}$$

for $t_2 \geq t_1$. Since $U_z(\underline{u})$ is a non-increasing function of z, and $Q(z,\ell)$ is the mean value of $U_z(\underline{u})$ in the interval $[z, z+\ell]$, we have

$$U_{z+\ell}(\underline{u}) \leq Q(z,\ell) \leq U_z(\underline{u})\ . \tag{1.95}$$

By (1.94) and (1.95) we obtain

$$U_{t_2+\ell}(\underline{u}) \leq e^{-(t_2-t_1)/k(\ell)}\,U_{t_1}(\underline{u})\ . \tag{1.96}$$

The inequality (1.96) implies (1.76). □

According to Toupin, the parameter $\ell > 0$ is to be chosen in a manner which will provide a small value for $k(\ell)$.

It follows from (1.76) that, given $\varepsilon > 0$ we have

$$\frac{U_z(\underline{u})}{U_0(\underline{u})} < \varepsilon\ ,$$

provided

$$z > \ell + k(\ell)\ln\frac{1}{\varepsilon}\ .$$

In [146], Toupin employs a mean value theorem due to Diaz and Payne [27], to obtain a pointwise estimate for the magnitude of the strain tensor at interior points of the cylinder. A similar estimate was established by Fichera [39] for an isotropic cylinder. We present here the estimate obtained in [39]. Let D be a bounded regular region.

<u>Lemma 1.1.</u> Let f be a biharmonic scalar field on D, and suppose that $f \in L^2(D)$. Let d be the distance of the point $\underline{x}$ of D from ∂D. Then the following estimate holds

$$|f(\underline{x})| \leq 1.9144\,d^{-3/2}\left(\int_D |f|^2\,dv\right)^{1/2}\ . \tag{1.97}$$

<u>Proof.</u> Let Ω be the ball of center $\underline{x}$ and unit radius. For each $\underline{y} \in D$ we set $\underline{y} = \underline{x} + r\underline{\zeta}$, where $\underline{\zeta} \in \partial\Omega$. In spherical coordinates the Laplacian operator

$$\Delta = \frac{\partial^2}{\partial y_i \partial y_i} ,$$

appears as

$$\Delta = \Delta_0 + \Delta_*, \ \Delta_0 = \frac{1}{r^2}\frac{\partial}{\partial r}\left(r^2 \frac{\partial}{\partial r}\right) ,$$

$$\Delta_* = \frac{1}{\sin\theta}\frac{\partial}{\partial\theta}\left(\sin\theta \frac{\partial}{\partial\theta}\right) + \frac{1}{\sin^2\theta}\frac{\partial^2}{\partial\varphi^2} .$$

In view of the relation

$$\int_{\partial\Omega} \Delta_* f(\underline{y}) da = 0 \qquad (da = \sin\theta \, d\theta d\varphi),$$

the equation

$$\Delta\Delta f = 0 ,$$

yields

$$\Delta_0 \Delta_0 \int_{\partial\Omega} f \, da = 0.$$

This equation implies that

$$\int_{\partial\Omega} f(\underline{y}) da = c_1 r^{-1} + c_2 + c_3 r + c_4 r^2 ,$$

where c_1, c_2, c_3, c_4 are real constants.

Since

$$\lim_{r \to 0} f(\underline{y}) = f(\underline{x}) , \qquad \lim_{r \to 0} r\Delta_0 f(\underline{y}) = 0 ,$$

uniformly with respect to $\underline{\zeta}$, we obtain $c_1 = c_3 = 0$, $c_2 = 4\pi f(\underline{x})$. Thus, we arrive at

$$\int_{\partial\Omega} f(\underline{y}) da = 4\pi f(\underline{x}) + r^2 c_4 .$$

If $\underline{y}^* = \underline{x} + \alpha r \underline{\zeta}$, with $0 < \alpha < 1$, then

$$\int_{\partial\Omega} f(\underline{y}^*) da = 4\pi f(\underline{x}) + \alpha^2 r^2 c_4 .$$

Thus, we obtain

$$f(\underline{x}) = \frac{1}{4\pi(1-\alpha^2)} \left(\int_{\partial\Omega} f(\underline{y}^*)da - \alpha^2 \int_{\partial\Omega} f(\underline{y})da \right) .$$

Multiplying the last equality by r^2 and integrating with respect to r from $r = 0$ to $r = d$, we find that

$$\frac{d^3}{3} f(\underline{x}) = \frac{1}{4\pi(1-\alpha^2)} \left(\frac{1}{\alpha^3} \int_{S(\alpha d)} f\, dv - \alpha^2 \int_{S(d)} f\, dv \right),$$

where $S(\rho)$ is the ball with radius ρ and center at $\underline{x}$.

If we apply the Schwarz inequality, we obtain

$$|f(\underline{x})| \leq g(\alpha) d^{-3/2} \left(\int_D |f|^2\, dv \right)^{1/2} ,$$

where

$$g(\alpha) = \frac{\sqrt{3}}{2\sqrt{\pi}} \frac{1+\alpha^{7/2}}{(1-\alpha^2)\alpha^{3/2}} , \quad 0 < \alpha < 1.$$

The function g attains an absolute minimum which is less than 1.9144. Thus, the last inequality implies (1.97). □

A direct proof of (1.97) can be obtained by the mean value theorem of Nicolesco [114]. The derivation used here follows that in [39].

If $\underline{u}$ is an equilibrium displacement field for a homogeneous and isotropic body, then the strain tensor $\underline{\underline{E}}(\underline{u})$ is biharmonic (cf. [47], Sect. 42).

Applying Lemma 1.1 to the function $\underline{\underline{E}}(\underline{u})$ on B_z, we obtain

$$|\{\underline{\underline{E}}(\underline{u})\}(\underline{x})| \leq 1.9144\, d^{-3/2} \left\{ \int_{B_z} |\underline{\underline{E}}(\underline{u})|^2 \right\}^{1/2} . \tag{1.98}$$

By (1.83) and (1.98),

$$|\{\underline{\underline{E}}(\underline{u})\}(\underline{x})| \leq 1.9144 \left\{ \frac{2}{\mu_m d^3} U_z(\underline{u}) \right\}^{1/2} .$$

When combined with energy inequality (1.97), the above inequality yields pointwise exponential decay for the magnitude of the strain tensor at interior points of the cylinder.

Pointwise estimates near the boundary have been obtained by Roseman [128] and Fichera [39]. Roseman established a pointwise estimate for the stress in a homogeneous and isotropic cylinder. When combined with Toupin's energy inequality this gives pointwise exponential decay for the stress throughout cylinder. Fichera presented a pointwise estimate for the strain tensor near any regular point of the boundary, under hypotheses on Γ which are less restrictive than those imposed in [128].

2. THEORY OF LOADED CYLINDERS: THE PROBLEMS OF ALMANSI AND MICHELL

2.1. Preliminaries

This chapter is concerned with the generalization of the relaxed Saint-Venant's problem to the case when the cylinder is subject to body forces and to surface tractions on the lateral boundary. This problem was initiated by Almansi [1] and Michell [102] and was developed in various later papers (see e.g. [139], [140], [12], [49]). The researches devoted to this subject are based on the semi-inverse method. In this chapter an adaptation of the method of [79] is used to provide a rational tool to solve the problem. The method offers a systematic approach which avoids artificial a priori assumptions.

We assume that a continuous body force field $\underline{f}$ is prescribed on B. By an equilibrium displacement field on B corresponding to the body force field $\underline{f}$ we mean a vector field $\underline{u} \in C^1(\bar{B}) \cap C^2(B)$ that satisfies the displacement equation of equilibrium

$$\operatorname{div} \underline{S}(\underline{u}) + \underline{f} = \underline{0} \quad \text{on B.} \tag{2.1}$$

We assume now that the boundary conditions (1.11) are replaced by

$$\underline{s}(\underline{u}) = \underline{p} \text{ on } \Pi, \quad \underline{R}(\underline{u}) = \underline{F}, \quad \underline{H}(\underline{u}) = \underline{M}, \tag{2.2}$$

where $\underline{p}$ is a vector-valued function preassigned on Π, and $\underline{F}$ and $\underline{M}$ are prescribed vectors. Suppose that $\underline{p}$ is piecewise regular on Π.

The problem consists in finding an equilibrium displacement field on B that corresponds to the body force field $\underline{f}$ and satisfies the boundary conditions (2.2).

When $\underline{f}$ and $\underline{p}$ are independent of the axial coordinate, the problem was first considered by Almansi [1] and Michell [102]. This particular case defines what is nowadays known in the literature as the Almansi-Michell problem.

Almansi [1] studied also the case when the prescribed forces are polynomials in the axial coordinate. This problem is known as the Almansi problem.

In this chapter we present a study of the foregoing problems.

We assume that the body is homogeneous and isotropic.

The next theorem will be of future use.

<u>Theorem 2.1.</u> If $\underline{u} \in C^1(\bar{B}) \cap C^2(B)$, then

$$\underline{R}(\underline{u}_{,3}) = \int_{\partial\Sigma_1} \underline{s}(\underline{u})ds - \int_{\Sigma_1} \operatorname{div} \underline{S}(\underline{u})da,$$

$$H_\alpha(\underline{u}_{,3}) = \int_{\partial\Sigma_1} \varepsilon_{\alpha\beta}x_\beta s_3(\underline{u})ds - \int_{\Sigma_1} \varepsilon_{\alpha\beta}x_\beta(S_{3j}(\underline{u}))_{,j}da + \varepsilon_{\alpha\beta}R_\beta(\underline{u}),$$

$$H_3(\underline{u}_{,3}) = \int_{\partial\Sigma_1} \varepsilon_{\alpha\beta}x_\alpha s_\beta(u)ds - \int_{\Sigma_1} \varepsilon_{\alpha\beta}x_\alpha(S_{\beta j}(\underline{u}))_{,j}\, da .$$

The proof of this theorem is strictly analogous to that given for Theorem 1.1 and can safely be omitted.

2.2. Almansi-Michell Problem

We assume throughout this section that $\underline{f}$ and $\underline{p}$ are independent of the axial coordinate.

We denote by (P_3) the Almansi-Michell problem corresponding to the system of loads $\{\underline{F},\underline{M},\underline{f},\underline{p}\}$. Let $K_{III}(\underline{F},\underline{M},\underline{f},\underline{p})$ denote the class of solutions to the problem (P_3). Recall that $K(\underline{F},\underline{M})$ denotes the class of solutions to the relaxed Saint-Venant's problem corresponding to the resultants $\underline{F}$ and $\underline{M}$.

Theorem 2.1 has the following important consequence:

<u>Corollary 2.1.</u> If $\underline{u} \in K_{III}(\underline{F},\underline{M},\underline{f},\underline{p})$ and $\underline{u}_{,3} \in C^1(\bar{B}) \cap C^2(B)$, then $\underline{u}_{,3} \in K(\underline{G},\underline{Q})$, where

$$\underline{G} = \int_\Gamma \underline{p}\, ds + \int_\Sigma \underline{f}\, da ,$$

$$Q_\alpha = \int_\Gamma \varepsilon_{\alpha\beta}x_\beta p_3 ds + \int_\Sigma \varepsilon_{\alpha\beta}x_\beta f_3 da + \varepsilon_{\alpha\beta}F_\beta , \tag{2.3}$$

$$Q_3 = \int_\Gamma \varepsilon_{\alpha\beta}x_\alpha p_\beta ds + \int_\Sigma \varepsilon_{\alpha\beta}x_\alpha f_\beta da .$$

Corollary 2.1 allows us to establish a method to derive a solution of the problem (P_3). Let $\underline{u}^0 \in K(\underline{G},\underline{Q})$ be Saint-Venant's solution. In view of Corollary 2.1, it is natural to enquire whether there exists a solution $\underline{u}'$ of the problem (P_3) such that $\underline{u}'_{,3} = \underline{u}^0$ modulo a rigid displacement. Thus, in view of Theorems 1.2, 1.4 and (1.37) we are led to seek a solution of the problem (P_3) in the form

$$\underline{u} = \int_0^{x_3}\int_0^{x_3} \underline{v}\{\hat{b}\}dx_3dx_3 + \int_0^{x_3} \underline{v}\{\hat{c}\}dx_3 + \underline{v}\{\hat{d}\} + x_3\underline{w}^0 + \underline{w}' , \tag{2.4}$$

where $\hat{b},\hat{c}$ and $\hat{d}$ are four-dimensional constant vectors, and $\underline{w}^0$ and $\underline{w}'$ are vector fields independent of x_3 such that $\underline{w}^0,\underline{w}' \in C^1(\bar{\Sigma})\cap C^2(\Sigma)$. Here $\underline{v}\{\hat{a}\}$ is defined by (1.27).

We assume for the remainder of this chapter that the body force and surface force belong to C^∞, and that the domain Σ is C^∞ smooth. We consider only a "C^∞-theory" but it is possible to get a solution of the problem for more general assumption of regularity. We have choosen these hypotheses in order to best emphasize the method for the solution of the problem.

Theorem 2.2. Let X be the set of all vector fields $\underline{u}$ of the form (2.4). Then there exists a vector field $\underline{u}' \in X$ which is solution of Almansi-Michell problem.

Proof. Let us prove that we can determine the vectors $\hat{b},\hat{c}$ and $\hat{d}$, and the vector fields $\underline{w}^0$ and $\underline{w}'$, such that $\underline{u} \in K_{III}(\underline{F},\underline{M},\underline{f},\underline{p})$. The determination of the unknown $\hat{b},\hat{c}$ and $\underline{w}^0$ is immediate. Thus, if $\underline{u}' \in X$ and $\underline{u}' \in K_{III}(\underline{F},\underline{M},\underline{f},\underline{p})$, then by Corollary 2.1, Theorem 1.2 and (2.4),

$$\int_0^{x_3} \underline{v}\{\hat{b}\}dx_3 + \underline{v}\{\hat{c}\} + \underline{w}^0 \in K(\underline{G},\underline{Q}) .$$

In view of Theorem 1.4, (1.39),(1.44) and (2.3) we conclude that

$$\begin{gathered} E(I_{\alpha\beta}b_\beta + Ax^0_\alpha b_3) = -\int_\Gamma p_\alpha ds - \int_\Sigma f_\alpha da, \\ b_\alpha x^0_\alpha + b_3 = 0 , \qquad b_4 = 0 , \end{gathered} \tag{2.5}$$

and $w^0_\alpha = 0$, $w^o_3 = \psi$ where ψ is characterized by

$$\Delta\psi = -2(b_\rho x_\rho + b_3) \quad \text{on} \quad \Sigma ,$$

$$\frac{\partial\psi}{\partial n} = b_\alpha \nu x_\rho (x_\alpha n_\rho - \frac{1}{2} n_\alpha x_\rho) + b_3 \nu x_\alpha n_\alpha \quad \text{on } \Gamma.$$

Moreover,

$$E(I_{\alpha\beta}c_\beta + Ax_\alpha^0 c_3) = -\int_\Gamma x_\alpha p_3 ds - \int_\Sigma x_\alpha f_3 da - F_\alpha ,$$

$$AE(c_\rho x_\rho^0 + c_3) = -\int_\Gamma p_3 ds - \int_\Sigma f_3 da , \tag{2.6}$$

$$\mu D c_4 = -\int_\Gamma \varepsilon_{\alpha\beta} x_\alpha p_\beta ds - \int_\Sigma \varepsilon_{\alpha\beta} x_\alpha [f_\beta + \mu(\psi_{,\beta} + \tfrac{1}{2}\nu b_\beta x_\rho x_\rho)] da.$$

It follows from (1.27) and (2.4) that

$$u'_\alpha = -\frac{1}{24} b_\alpha x^4 - \frac{1}{6} c_\alpha x_3^3 - \frac{1}{2} d_\alpha x_3^2 - \frac{1}{2} c_4 \varepsilon_{\alpha\beta} x_\beta x_3^2 -$$

$$- d_4 \varepsilon_{\alpha\beta} x_\beta x_3 + \sum_{j=1}^{3} (d_j + c_j x_3 + \tfrac{1}{2} b_j x_3^2) w_\alpha^{(j)} + w'_\alpha , \tag{2.7}$$

$$u'_3 = \frac{1}{6} (b_\rho x_\rho + b_3) x_3^3 + \frac{1}{2} (c_\rho x_\rho + c_3) x_3^2 +$$

$$+ (d_\rho x_\rho + d_3) x_3 + c_4 x_3 \varphi + d_4 \varphi + x_3 \psi + \chi ,$$

where we have used the notation $w'_3 = \chi$.

The stress-displacement relations imply

$$S_{\alpha\beta}(\underline{u}') = T_{\alpha\beta}(\underline{w}') + \lambda(c_4\varphi + \psi)\delta_{\alpha\beta} ,$$

$$S_{\alpha 3}(\underline{u}') = \mu[(c_4 x_3 + d_4)(\varphi_{,\alpha} - \varepsilon_{\alpha\beta} x_\beta) + x_3 \psi_{,\alpha} + \chi_{,\alpha} +$$

$$+ \sum_{j=1}^{3} (c_j + b_j x_3) w_\alpha^{(j)} , \tag{2.8}$$

$$S_{33}(\underline{u}') = E[\tfrac{1}{2} (b_\rho x_\rho + b_3) x_3^2 + (c_\rho x_\rho + c_3) x_3 + d_\rho x_\rho + d_3] +$$

$$+ (\lambda + 2\mu)(c_4\varphi + \psi) + \lambda E_{\rho\rho}(\underline{w}'),$$

where

$$T_{\alpha\beta}(\underline{u}') = 2\mu E_{\alpha\beta}(\underline{u}') + \lambda \delta_{\alpha\beta} E_{\rho\rho}(\underline{w}') .$$

The first two equations of equilibrium and the first two conditions on lateral boundary become

$$(T_{\alpha\beta}(\underline{w}'))_{,\beta} + g_\alpha = 0 \quad \text{on } \Sigma, \quad T_{\alpha\beta}(\underline{w}')n_\beta = q_\alpha \quad \text{on } \Gamma, \tag{2.9}$$

where

$$\begin{aligned} g_\alpha &= f_\alpha + \lambda(c_4\varphi + \psi)_{,\alpha} + S_{\alpha 3}(\underline{u}'_{,3}), \\ q_\alpha &= p_\alpha - \lambda(c_4\varphi + \psi)n_\alpha . \end{aligned} \tag{2.10}$$

Thus, from (2.9) and (2.10) we conclude that $\{w'_\alpha, T_{\alpha\beta}(\underline{w}')\}$ is a plane elastic state corresponding to the body forces g_α and to the surface forces q_α. The necessary and sufficient conditions to solve the boundary-value problem (2.9) are

$$\int_\Sigma g_\alpha da + \int_\Gamma q_\alpha ds = 0, \quad \int_\Sigma \varepsilon_{\alpha\beta} x_\alpha g_\beta da + \int_\Gamma \varepsilon_{\alpha\beta} x_\alpha q_\beta ds = 0. \tag{2.11}$$

By the divergence theorem and (2.10),

$$\int_\Sigma g_\alpha da + \int_\Gamma q_\alpha ds = -R_\alpha(\underline{u}'_{,3}) + \int_\Sigma f_\alpha da + \int_\Gamma p_\alpha ds,$$

$$\int_\Sigma \varepsilon_{\alpha\beta} x_\alpha g_\beta da + \int_\Gamma \varepsilon_{\alpha\beta} x_\alpha q_\beta ds = -H_3(\underline{u}'_{,3}) + \int_\Sigma \varepsilon_{\alpha\beta} x_\alpha f_\beta da + \int_\Gamma \varepsilon_{\alpha\beta} x_\alpha p_\beta ds.$$

Note that the conditions

$$R_\alpha(\underline{u}'_{,3}) = G_\alpha, \qquad H_3(\underline{u}'_{,3}) = Q_3 ,$$

were used to obtain (2.5) and (2.6). Thus we conclude that the conditions (2.11) are satisfied on the basis of (2.5) and (2.6). In what follows we assume that the functions w'_α are known.

The last equation of equilibrium and the last condition on the lateral boundary become

$$\begin{aligned} \mu\Delta\chi &= -f_3 - 2\mu(c_\varrho x_\varrho + c_3) \quad \text{on } \Sigma, \\ \mu\frac{\partial\chi}{\partial n} &= p_3 + \mu\nu n_\alpha x_\alpha(c_\varrho x_\varrho + c_3) - \frac{1}{2} c_\alpha \mu\nu n_\alpha x_\varrho x_\varrho \quad \text{on } \Gamma, \end{aligned} \tag{2.12}$$

respectively. The necessary and sufficient condition to solve the boundary-value problem (2.12) is satisfied on the basis of $(2.6)_2$. In what follows we assume that χ is known.

Let us note that on the basis of (2.6) and (2.8) we have

$$S_{33}(\underline{u}'_{,3}) = E[(b_\varrho x_\varrho + b_3)x_3 + c_\varrho x_\varrho + c_3],$$
$$H_\alpha(\underline{u}'_{,3}) = -\varepsilon_{\alpha\beta}E(I_{\beta\varrho}c_\varrho + Ax^0_\beta c_3) = \varepsilon_{\alpha\beta}F_\beta + \varepsilon_{\alpha\beta}(\int_\Gamma x_\beta p_3 ds + \int_\Sigma x_\beta f_3 da). \tag{2.13}$$

In view of Theorem 2.1,

$$H_\alpha(\underline{u}'_{,3}) = \varepsilon_{\alpha\beta}(\int_\Gamma x_\beta p_3 ds + \int_\Sigma x_\beta f_3 da) + \varepsilon_{\alpha\beta}R_\beta(\underline{u}') . \tag{2.14}$$

It follows from (2.13) and (2.14) that $R_\alpha(\underline{u}') = F_\alpha$.

The conditions $R_3(\underline{u}') = F_3$, $\underline{H}(\underline{u}) = \underline{M}$ reduce to

$$E(I_{\alpha\beta}d_\beta + Ax^0_\alpha d_3) = \varepsilon_{\alpha\beta}M_\beta - \int_\Sigma x_\alpha N\, da,$$
$$EA(d_\alpha x^0_\alpha + d_3) = -F_3 - \int_\Sigma N\, da , \tag{2.15}$$
$$\mu D d_4 = -M_3 - \mu\int_\Sigma \varepsilon_{\alpha\beta}x_\alpha(\chi_{,\beta} + \sum_{i=1}^{3} c_i\, w^{(i)}_\beta) da ,$$

where

$$N = \lambda E_{\varrho\varrho}(\underline{w}') + (\lambda + 2\mu)(c_4\varphi + \psi).$$

The system (2.15) can always be solved for d_1, d_2, d_3 and d_4.

Thus, if $\hat{b}, \hat{c}$ and $\hat{d}$ are given by (2.5),(2.6) and (2.15), and the functions w'_i are characterized by (2.9) and (2.12), then $\underline{u}' \in K_{III}(\underline{F},\underline{M},\underline{f},\underline{p})$. □

2.3. Almansi's Problem

In this section we assume that the body force field $\underline{f}$ and the surface force $\underline{p}$ are polynomials of degree r in the axial coordinate, i.e.

$$\underline{f} = \sum_{k=0}^{r} \underline{f}_k\, x_3^k \quad , \quad \underline{p} = \sum_{k=0}^{r} \underline{p}_k\, x_3^k , \tag{2.16}$$

where $\underline{f}_k$ and $\underline{p}_k$ are independent of x_3.

Almansi's problem consists in finding an equilibrium displacement field on B that corresponds to the body force field $\underline{f}$ and satisfies the boundary conditions (2.2), when $\underline{f}$ and $\underline{p}$ are given by (2.16). We refer to this problem as the problem (A).

Let (A_o) denote the Almansi-Michell problem corresponding to the system of loads $\{\underline{F},\underline{M},\underline{f} = \underline{f}_o, \underline{p} = \underline{p}_o\}$. We denote by (A_n) $(n=1,2,\dots,r)$ the Almansi problem corresponding to the system of loads $\{\underline{F} = \underline{O}, \underline{M} = \underline{O}, \underline{f} = \underline{f}_n x_3^n, \underline{p} = \underline{p}_n x_3^n\}$. Let $\underline{u}^{(n)}$ be a solution of the problem (A_n) $(n=0,1,2,\dots,r)$. Since the mathematical theory under consideration is linear, the vector field $\underline{u}$ defined by

$$\underline{u} = \sum_{k=0}^{r} \underline{u}^{(k)} x_3^k , \tag{2.17}$$

is a solution of the problem (A).

In what follows we use the method of induction to establish a solution of the problem (A). In the previous section we obtained a solution of the problem (A_o). Our task is to establish a solution of the problem (A_{n+1}) once a solution of the problem (A_n) (with $\underline{F} = \underline{O}$, $\underline{M} = \underline{O}$) is known.

Let $Q_n(\underline{f}_n x_3^n, \underline{p}_n x_3^n)$ $(n=0,1,2,\dots,r)$ be the class of solutions to the Almansi problem corresponding to the system of loads $\{\underline{F} = \underline{O}, \underline{M} = \underline{O}, \underline{f} = \underline{f}_n x_3^n, \underline{p} = \underline{p}_n x_3^n\}$.

By induction hypothesis, we know to derive a solution $\hat{\underline{u}} \in Q_n(\underline{f}_n x_3^n, \underline{p}_n x_3^n)$. It follows that we also know a solution $\underline{u}^* \in Q_n(\underline{f}_{n+1} x_3^n, \underline{p}_{n+1} x_3^n)$. Thus we are led to the following problem: to find a vector field $\underline{u}'' \in Q_{n+1}(\underline{f}_{n+1} x_3^{n+1}, \underline{p}_{n+1} x_3^{n+1})$ when $\underline{u}^* \in Q_n(\underline{f}_{n+1} x_3^n, \underline{p}_{n+1} x_3^n)$ is given. We refer to this problem as the problem (D).

In order to solve this problem we need the following

<u>Lemma 2.1.</u> If $\underline{u} \in Q_{n+1}(\underline{f}_{n+1} x_3^{n+1}, \underline{p}_{n+1} x_3^{n+1})$ and $\underline{u}_{,3} \in C^1(\bar{B}) \cap C^2(B)$, then

$$(n+1)^{-1} \underline{u}_{,3} \in Q_n(\underline{f}_{n+1} x_3^n, \underline{p}_{n+1} x_3^n).$$

<u>Proof.</u> Let $\underline{u} \in Q_{n+1}(\underline{f}_{n+1} x_3^{n+1}, \underline{p}_{n+1} x_3^{n+1})$ such that $\underline{u}_{,3} \in C^1(\bar{B}) \cap C^2(B)$. It follows from (2.1) and (2.2) that

$$\operatorname{div} \underline{S}(\underline{u}_{,3}) + (n+1)\underline{f}_{n+1} x_3^n = 0 \quad \text{on } B,$$
$$\underline{s}(\underline{u}_{,3}) = (n+1)\underline{p}_{n+1} x_3^n \qquad \text{on } \Pi.$$

Since the theory under consideration is linear, the vector field $\underline{u}' = (n+1)^{-1}\underline{u}_{,3}$ is an equilibrium displacement field on B that corresponds to the body force field $\underline{f}_{n+1}x_3^n$ and satisfies the condition $\underline{s}(\underline{u}') = \underline{p}_{n+1}x_3^n$ on Π. In view of Theorem 2.1 we find $\underline{R}(\underline{u}') = \underline{0}$, $\underline{H}(\underline{u}') = \underline{0}$. This completes the proof of the lemma. □

Lemma 2.1 allows us to solve the problem (D). Thus, in view of this lemma we are led to seek the vector field $\underline{u}''$ such that $(n+1)^{-1}\underline{u}''_{,3} = \underline{u}^*$ modulo a rigid displacement, i.e.

$$(n+1)^{-1}\underline{u}''_{,3} = \underline{u}^* + \underline{\alpha} + \underline{\beta} \times \underline{x} , \qquad (2.18)$$

where $\underline{\alpha}$ and $\underline{\beta}$ are constant vectors. Then it follows that

$$\begin{aligned} u''_\alpha &= (n+1)\Big[\int_0^{x_3} u^*_\alpha dx_3 - \frac{1}{2} a_\alpha x_3^2 - a_4 \varepsilon_{\alpha\beta} x_\beta x_3 + w^*_\alpha(x_1,x_2)\Big], \\ u''_3 &= (n+1)\Big[\int_0^{x_3} u^*_3 dx_3 + (a_\rho x_\rho + a_3)x_3 + w^*_3(x_1,x_2)\Big], \end{aligned} \qquad (2.19)$$

except for an additive rigid displacement. Here $\underline{w}^*$ is an arbitrary vector field independent of x_3, and we have used the notations $a_\alpha = \varepsilon_{\rho\alpha}\beta_\rho$, $a_3 = \alpha_3$, $a_4 = \beta_3$.

Theorem 2.3. Let $\underline{u}^* \in Q_n(\underline{f}_{n+1}x_3^n, \underline{p}_{n+1}x_3^n)$, and let Y be the set of all vector fields of the form (2.19). Then there exists a vector field $\underline{u}'' \in Y$ such that $\underline{u}'' \in Q_{n+1}(\underline{f}_{n+1}x_3^{n+1}, \underline{p}_{n+1}x_3^{n+1})$.

Proof. We must prove that the functions w_i^* and the constants a_s (s= =1,2,3,4) can be determined so that $\underline{u}'' \in Q_{n+1}(\underline{f}_{n+1}x_3^{n+1}, \underline{p}_{n+1}x_3^{n+1})$.

For convenience, we introduce the vector field $\underline{w}$ by

$$w^*_\alpha = \sum_{i=1}^{3} a_i \underline{w}^{(i)}_\alpha + w_\alpha , \quad w_3 = a_4\varphi + w_3 ,$$

where the functions $w^{(i)}_\alpha$ are given by (1.25), and the function φ is characterized by (1.26).

It follows from (2.19) that

$$u_1'' = (n+1)\Big[\int_0^{x_3} u_1^* dx_3 - \tfrac{1}{2}\, a_1 x_3^2 - a_4 x_2 x_3 - \tfrac{1}{2}\, a_1 \nu (x_1^2 - x_2^2) - $$

$$- a_2 \nu x_1 x_2 - a_3 \nu x_1 + w_1\Big],$$

$$u_2'' = (n+1)\Big[\int_0^{x_3} u_2^* dx_3 - \tfrac{1}{2}\, a_2 x_3^2 + a_4 x_1 x_3 - a_1 \nu x_1 x_2 - \qquad (2.20)$$

$$- \tfrac{1}{2}\, a_2 \nu (x_2^2 - x_1^2) - a_3 \nu x_2 + w_2\Big],$$

$$u_3'' = (n+1)\Big[\int_0^{x_3} u_3^* dx_3 + (a_\rho x_\rho + a_3) x_3 + a_4 \varphi + w_3\Big] .$$

The stress-displacement relations imply

$$S_{\alpha\beta}(\underline{u}'') = (n+1)\Big[\int_0^{x_3} S_{\alpha\beta}(\underline{u}^*) dx_3 + T_{\alpha\beta}(\underline{w}) + \lambda \delta_{\alpha\beta} u_3^*(x_1, x_2, 0)\Big],$$

$$S_{33}(\underline{u}'') = (n+1)\Big[\int_0^{x_3} S_{33}(\underline{u}^*) dx_3 + E(a_\rho x_\rho + a_3) + \lambda w_{\rho,\rho} + (\lambda + 2\mu) u_3^*(x_1, x_2, 0)\Big], \quad (2.21)$$

$$S_{\alpha 3}(\underline{u}'') = (n+1)\Big[\int_0^{x_3} S_{\alpha 3}(\underline{u}^*) dx_3 + \mu a_4 (\varphi_{,\alpha} - \varepsilon_{\alpha\beta} x_\beta) + w_{3,\alpha} + \mu u_\alpha^*(x_1, x_2, 0)\Big].$$

Clearly,

$$(S_{\alpha i}(\underline{u}''))_{,i} = (n+1)\Big[\int_0^{x_3} (S_{\alpha i}(\underline{u}^*))_{,i} dx_3 + (T_{\alpha\beta}(\underline{w}))_{,\beta} + g_\alpha\Big],$$

$$(S_{3i}(\underline{u}''))_{,i} = (n+1)\Big[\int_0^{x_3} (S_{3i}(\underline{u}^*))_{,i} dx_3 + \mu \Delta w_3 + g\Big], \qquad (2.22)$$

where

$$g_\alpha = (S_{3\alpha}(\underline{u}^*))(x_1, x_2, 0) + \lambda u_{3,\alpha}^*(x_1, x_2, 0),$$

$$g = (S_{33}(\underline{u}^*))(x_1, x_2, 0) + \mu\, u_{\alpha,\alpha}^*(x_1, x_2, 0) . \qquad (2.23)$$

Since $\underline{u}^* \in Q_n(\underline{f}_{n+1} x_3^n , \underline{p}_{n+1} x_3^n)$, the equations of equilibrium and the condition on the lateral boundary reduce to

$$(T_{\alpha\beta}(\underline{w}))_{,\beta} + g_\alpha = 0 \text{ on } \Sigma, \quad T_{\alpha\beta}(\underline{w}) n_\beta = q_\alpha \text{ on } \Gamma, \qquad (2.24)$$

$$\mu \Delta w_3 + g = 0 \text{ on } \Sigma, \quad \mu \frac{\partial w_3}{\partial n} = q \text{ on } \Gamma, \tag{2.25}$$

where

$$q_\alpha = -\lambda n_\alpha u_3^*(x_1,x_2,0), \quad q = -\mu n_\alpha u_\alpha^*(x_1,x_2,0).$$

It follows from (2.24) that $\{w_\alpha, T_{\alpha\beta}(\underline{w})\}$ is a plane elastic state corresponding to the body forces g_α and to the surface forces q_α. It is a simple matter to see that the necessary and sufficient conditions to solve the plane strain problem (2.24) are satisfied.

The function w_3 is characterized by (2.25).

In view of Theorem 2.1 we find that $R_\alpha(\underline{u}'') = \varepsilon_{\beta\alpha} H_\beta((n+1)\underline{u}^*) = 0$.

The conditions $R_3(\underline{u}'') = 0$, $\underline{H}(\underline{u}'') = \underline{0}$ reduce to

$$E(I_{\alpha\beta} a_\beta + A x_\alpha^0 a_3) = -\int_\Sigma x_\alpha [\lambda w_{\rho,\rho} + (\lambda+2\mu) u_3^*(x_1,x_2,0)]\, da,$$

$$AE(a_\rho x_\rho^0 + a_3) = -\int_\Sigma [\lambda w_{\rho,\rho} + (\lambda+2\mu) u_3^*(x_1,x_2,0)]\, da, \tag{2.26}$$

$$D a_4 = -\int_\Sigma \varepsilon_{\alpha\beta} x_\alpha [w_{3,\beta} + u_\beta^*(x_1,x_2,0)]\, da.$$

The system (2.26) can always be solved for a_1, a_2, a_3 and a_4. □

Remark. It follows from (1.27),(2.19) and (2.20) that the solution $\underline{u}''$ may be written in the form

$$\underline{u}'' = (n+1)\left[\int_0^{x_3} \underline{u}^* dx_3 + \underline{v}\{\hat{a}\} + \underline{w}\right]. \tag{2.27}$$

Here w_α are the components of the displacement field in the plane strain problem (2.24), w_3 is characterized by (2.25) and $\hat{a}$ is determined by (2.26).

The above results yield a rational scheme to derive a solution to the Almansi problem.

2.4. Applications

In this section we use the results established in the preceding two paragraphs in order to study the problem of thermal stresses in homogeneous and isotropic cylinders within the linear theory of thermoelastostatics.

Let T be the absolute temperature measured from the constant absolute temperature in the reference configuration. In the equilibrium theory of linear thermoelasticity, the temperature field T can be found by solving the heat flow boundary-value problem associated with the heat conduction and energy equations. In this section we shall treat the temperature field T as having already been so determined.

As is usual in thermoelastostatics, we assume that mechanical loads are absent. Thus, the principal attention is devoted to the deformation due to the temperature field.

We consider a relaxed formulation of the problem in which the detailed assignment of the terminal tractions is abandoned in favor of prescribing merely the approapriate stress resultants.

According to the body force analogy (cf.[16], Sect.11), the thermoelastic problem reduces to the problem of finding an equilibrium displacement field $\underline{u}$ on B that corresponds to the body force field $\underline{f} = -\beta \operatorname{grad} T$ and satisfies the conditions

$$\underline{s}(\underline{u}) = \underline{p} \text{ on } \Pi,\quad R_\alpha(\underline{u}) = 0,\quad R_3(\underline{u}) = -\int_{\Sigma_1} \beta T\, da,$$
$$H_\alpha(\underline{u}) = -\int_{\Sigma_1} \beta \varepsilon_{\alpha\rho} x_\rho T\, da,\quad H_3(\underline{u}) = 0\ , \tag{2.28}$$

where $\underline{p} = \beta T\, \underline{n}$. Here β is the stress-temperature modulus. We refer to the foregoing problem as the problem (Z).

i) <u>Plane temperature field.</u> We now consider the case when the temperature field is independent of the axial coordinate, i.e.

$$T = T_0(x_1,x_2) \qquad (x_1,x_2) \in \Sigma\ ,$$

where $T_0 \in C^1(\bar{\Sigma}) \cap C^2(\Sigma)$ is a prescribed field.
Clearly, in this case the problem (Z) reduces to the Almansi-Michell problem which consists in finding a vector field $\underline{u} \in K_{III}(\underline{F},\underline{M},\underline{f},\underline{p})$ where

$$F_\alpha = 0,\ F_3 = -\int_{\Sigma} \beta T_0\, da,\ M_\alpha = -\int_{\Sigma} \beta \varepsilon_{\alpha\rho} x_\rho T_0\, da\ ,$$
$$M_3 = 0,\ f_\alpha = -\beta T_{0,\alpha}\ ,\ f_3 = 0,\ p_\alpha = \beta T_0 n_\alpha,\ p_3 = 0. \tag{2.29}$$

A solution of this problem is given by (2.7). In view of (2.29),

$$\int_{\Sigma} f_\alpha da + \int_{\Gamma} p_\alpha ds = 0 , \quad \int_{\Sigma} \varepsilon_{\alpha\beta} x_\alpha f_\beta da + \int_{\Gamma} \varepsilon_{\alpha\beta} x_\alpha p_\beta ds = 0. \tag{2.30}$$

It follows from (2.5) and (2.30) that $b_i = 0$. Clearly, $\psi = 0$ is a solution of the boundary-value problem (1.42). Then, it follows from (2.6) and (2.30) that $c_s = 0$ (s=1,2,3,4). Now we can see that $\chi = 0$ is a solution of the boundary-value problem (2.12). The functions w'_α are characterized by the plane strain problem

$$(T_{\alpha\beta}(\underline{w}'))_{,\beta} + f_\alpha = 0 \text{ on } \Sigma, \quad T_{\alpha\beta}(\underline{w}'))n_\beta = p_\alpha \text{ on } \Gamma , \tag{2.31}$$

where f_α and p_α are given by (2.29). The system (2.15) reduces to

$$E(I_{\alpha\beta} d_\beta + Ax^0_\alpha d_3) = \int_{\Sigma} \beta x_\alpha T_0 da - \lambda \int_{\Sigma} x_\alpha w'_{\rho,\rho} da ,$$

$$AE(d_\alpha x^0_\alpha + d_3) = \int_{\Sigma} \beta T_0 da - \lambda \int_{\Sigma} w'_{\rho,\rho} da, \quad d_4 = 0. \tag{2.32}$$

Thus we conclude that a solution of the problem is given by

$$u_1 = -\frac{1}{2} d_1 x_3^2 - \frac{1}{2} d_1 \nu (x_1^2 - x_2^2) - d_2 \nu x_1 x_2 - d_3 \nu x_1 + w'_1 ,$$

$$u_2 = -\frac{1}{2} d_2 x_3^2 - d_1 \nu x_1 x_2 - \frac{1}{2} d_2 \nu (x_1^2 - x_2^2) - d_3 \nu x_2 + w'_2 , \tag{2.33}$$

$$u_3 = (d_\rho x_\rho + d_3) x_3 .$$

If $T = T^* =$ constant, then

$$w'_\alpha = \frac{\beta}{2(\lambda+\mu)} T^* x_\alpha . \tag{2.34}$$

Let us suppose that the rectangular Cartesian coordinate frame is chosen in such a way that the origin O coincides with the centroid of Σ_1. Then, it follows from (2.32) and (2.34) that

$$d_\alpha = 0, \quad d_3 = \beta T^* / (3\lambda + 2\mu). \tag{2.35}$$

The solution is given by (2.33), (2.34) and (2.35).

ii) <u>The deformation of a cylinder under the action of a temperature field that is a polynomial in the axial coordinate.</u> We assume now that the temperature field is a polynomial of degree r in the axial coordinate, namely

$$T = \sum_{k=0}^{r} T_k x_3^k , \tag{2.36}$$

where T_k are independent of x_3.

In this case the problem (Z) reduces to the Almansi problem. We denote by (Z_n) $(n=0,1,2,\ldots,r)$ the problem (Z) corresponding to the temperature field $T = T_n x_3^n$. Clearly, if we know the solution of the problem (Z_n), for any n, then we can establish a solution of the problem (Z) for the temperature distribution (2.36). The solution of the problem (Z_0) has been established previously. We must derive the solution $\underline{u}''$ of the problem (Z_{n+1}) when the solution of the problem (Z_n) is known. As the solution of the problem (Z_n) is known for any T_n, it follows that we know the solution $\underline{u}^*$ of the problem corresponding to the temperature field $T = T_{n+1}x_3^n$. According to Theorem 2.3 the vector field $\underline{u}''$ is given by (2.27), where $\underline{w}$ is characterized by (2.24) and (2.25), and $\hat{a}$ is determined by (2.26).

If the temperature field is linear in x_3,

$$T = T_0 + T_1 x_3 ,$$

where T_0 and T_1 are prescribed constants, then a simple calculation shows that

$$u_\alpha = \frac{\beta}{3\lambda + 2\mu} x_\alpha (T_0 + T_1 x_3) ,$$

$$u_3 = \frac{\beta}{3\lambda + 2\mu}\left[(T_0 + \frac{1}{2} T_1 x_3)x_3 - \frac{1}{2} T_1 x_\rho x_\rho\right].$$

Other results concerning the problem of thermal stresses in elastic cylinders have been presented in [62],[73] and [74].

3. ANISOTROPIC MATERIALS

3.1. Preliminaries

The relaxed Saint-Venant's problem for anisotropic elastic bodies has been extensively studied (see, for example, Lekhnitskii [96] and Hatiashvili [49]).

A comprehensive bibliography of the vast literature concerned to this subject would multiply the length of this study. We note that the researches devoted to the relaxed Saint-Venant's problem are based on various assumptions regarding the structure of the prevailing fields of displacement or stress. It is the purpose of this chapter to extend the results derived in the previous chapters to the case of anisotropic elastic bodies with general elasticities. The procedure presented in this chapter avoids the semi-inverse method and permits a treatment of the problem even for nonhomogeneous bodies, where the elasticity tensor is independent of the axial coordinate. Saint-Venant's problem for nonhomogeneous elastic cylinders where the elastic coefficients are independent of the axial coordinate has been studied in various papers (see, e.g. [147],[126], [135],[141],[97],[76]). According to Toupin [146], the proof of Saint-Venant's principle presented in Section 1.6 also remains valid for this kind of nonhomogeneous elastic bodies.

In the first part of the chapter we present in a concise and rational form a solution to the relaxed Saint-Venant's problem for anisotropic elastic bodies. This solution coincides with that given in [67] and incorporates the solutions presented in [96],[12],[49]. An illustrative application is given. Then, minimum energy characterizations of the solutions are established. The results of Day [25] and Podio-Guidugli [123] are extended to study Truesdell's problem for anisotropic elastic cylinders. Also included in this chapter is a study of the problems of Almansi and Michell.

We assume for the remainder of this chapter that the field $\underline{C}$ is independent of the axial coordinate, i.e.

$$\underline{C} = \underline{C}(x_1,x_2) \qquad (x_1,x_2) \in \Sigma \quad . \tag{3.1}$$

Let $\mathfrak{D}^*$ denote the set of all equilibrium displacement fields $\underline{u}$ that satisfy the condition $\underline{s}(\underline{u}) = \underline{0}$ on the lateral boundary. For future use we record the following

Theorem 3.1. If $\underline{u} \in \mathfrak{D}^*$ and $\underline{u}_{,3} \in C^1(\bar{B}) \cap C^2(B)$, then $\underline{u}_{,3} \in \mathfrak{D}^*$ and

$$\underline{R}(\underline{u}_{,3}) = 0\ , \quad H_\alpha(\underline{u}_{,3}) = \varepsilon_{\alpha\beta} R_\beta(\underline{u}), \quad H_3(\underline{u}_{,3}) = 0\ .$$

The proof of this theorem, which we omit, is analogous to that given for Theorem 1.1.

Theorem 3.1 has the following consequences:

Corollary 3.1. If $\underline{u} \in K_I(F_3, M_1, M_2, M_3)$ and $\underline{u}_{,3} \in C^1(\bar{B}) \cap C^2(B)$, then $\underline{u}_{,3} \in \mathfrak{D}^*$ and

$$\underline{R}(\underline{u}_{,3}) = \underline{0}\ , \quad \underline{H}(\underline{u}_{,3}) = \underline{0}\ .$$

Corollary 3.2. If $\underline{u} \in K_{II}(F_1, F_2)$ and $\underline{u}_{,3} \in C^1(\bar{B}) \cap C^2(B)$ then

$$\underline{u}_{,3} \in K_I(0, F_2, -F_1, 0).$$

3.2. Generalized Plane Strain Problem

The state of generalized plane strain of the cylinder B is characterized by

$$\underline{u} = \underline{u}(x_1, x_2) \qquad (x_1, x_2) \in \Sigma\ . \tag{3.2}$$

This restriction, in conjunction with the stress-displacement relation, implies that $\underline{S} = \underline{S}(x_1, x_2)$. Further,

$$S_{i\alpha}(\underline{u}) = C_{i\alpha k\beta} u_{k,\beta}\ . \tag{3.3}$$

A vector field $\underline{u}$ is an admissible displacement field provided

(i) $\underline{u}$ is independent of x_3;

(ii) $\underline{u} \in C^1(\bar{\Sigma}) \cap C^2(\Sigma)$.

Given body force $\underline{f}$ on B and surface force $\underline{p}$ on Π, with $\underline{f}$ and $\underline{p}$

independent of x_3, the generalized plane strain problem consists in finding an admissible displacement field $\underline{u}$ which satisfies the equations of equilibrium

$$(S_{i\alpha}(\underline{u}))_{,\alpha} + f_i = 0 \text{ on } \Sigma , \tag{3.4}$$

and the boundary conditions

$$S_{i\alpha}(\underline{u})n_\alpha = p_i \quad \text{ on } \Gamma . \tag{3.5}$$

The stress $S_{33}(\underline{u})$ can be determined after the displacement field $\underline{u}$ is found.

The generalized plane strain problem for homogeneous bodies was studied in various papers (see, for example, Lekhnitskii [96]).

The conditions of equilibrium for the cylinder B are equivalent to

$$\int_\Sigma \underline{f}da + \int_\Gamma \underline{p}ds = \underline{0}, \quad \int_\Sigma \varepsilon_{\alpha\beta}x_\alpha f_\beta da + \int_\Gamma \varepsilon_{\alpha\beta}x_\alpha p_\beta ds = 0, \tag{3.6}$$

$$\int_\Sigma x_\alpha f_3 da + \int_\Gamma x_\alpha p_3 ds = \int_\Sigma S_{3\alpha}(\underline{u})da . \tag{3.7}$$

It follows from (3.4),(3.5) and the divergence theorem that

$$\int_\Sigma S_{3\alpha}(\underline{u})da = \int_\Sigma \{S_{3\alpha}(\underline{u}) + x_\alpha[(S_{3\rho}(\underline{u}))_{,\rho} + f_3]\}da =$$

$$= \int_\Sigma [(x_\alpha S_{3\rho}(\underline{u}))_{,\rho} + x_\alpha f_3]da = \int_\Gamma x_\alpha p_3 ds + \int_\Sigma x_\alpha f_3 da.$$

Thus, the conditions (3.7) are identically satisfied.

We assume for the remainder of this chapter that $\underline{C} \in C^\infty(\bar{\Sigma})$ and that the domain Σ is C^∞-smooth. Moreover, we assume that $\underline{f}$ and $\underline{p}$ belong to C^∞.

Let $\mathcal{P}$ be the set of all admissible displacement fields, and let $\underline{L}$ be the operator on $\mathcal{P}$ defined by

$$L_i\underline{u} = -(C_{i\alpha k\beta}u_{k,\beta})_{,\alpha} .$$

The equations of equilibrium (3.4) take the form

$$\underline{L}\underline{u} = \underline{f} \text{ on } \Sigma . \tag{3.8}$$

The conditions (3.5) can be rewritten as

$$\underline{s}(\underline{u}) = \underline{p} \quad \text{on} \quad \Gamma . \tag{3.9}$$

Let $\underline{u},\underline{v} \in \mathcal{P}$. By the divergence theorem,

$$\int_{\Sigma} (L\underline{u})\cdot \underline{v}\, da = 2\int_{\Sigma} W(\underline{u},\underline{v})da - \int_{\Gamma} \underline{s}(\underline{u})\cdot \underline{v}\, ds , \tag{3.10}$$

where

$$2W(\underline{u},\underline{v}) = C_{i\alpha k\beta}\, E_{i\alpha}(\underline{u})E_{k\beta}(\underline{v}),$$

is the bilinear form corresponding to the quadratic form

$$2W(\underline{u}) = C_{i\alpha k\beta}\, E_{i\alpha}(\underline{u})\; E_{k\beta}(\underline{u}) .$$

Let $\underline{u}^*$ be a solution of the boundary-value problem (3.8),(3.9) corresponding to $\underline{f} = \underline{0}$ and $\underline{p} = \underline{0}$. Since $W(\underline{u})$ is positive definite in the variables $E_{s\beta}(\underline{u})$, it follows from (3.10) that

$$u^*_{\alpha} = a_{\alpha} + \varepsilon_{\alpha\beta} b\, x_{\beta} \;,\; u^*_3 = a_3 \;, \tag{3.11}$$

where a_i and b are arbitrary constants. Let us consider the boundary condition

$$\underline{s}(\underline{u}) = \underline{0} \;\text{on}\; \Gamma . \tag{3.12}$$

Following Fichera [37], a C^{∞} solution in $\bar{\Sigma}$ of the boundary-value problem (3.8),(3.12) exists if and only if

$$\int_{\Sigma} \underline{f}\cdot\underline{u}^*\, da = 0,$$

for any displacement field $\underline{u}^*$ given by (3.11). Thus, we are led to the

Theorem 3.2. Let $\underline{f}$ be a vector field of class C^{∞} on $\bar{\Sigma}$. The boundary-value problem (3.8),(3.12) has solutions belonging to $C^{\infty}(\bar{\Sigma})$ if and only if

$$\int_{\Sigma} \underline{f}\, da = \underline{0} \;,\; \int_{\Sigma} \varepsilon_{\alpha\beta} x_{\alpha} f_{\beta}\, da = 0. \tag{3.13}$$

It is a simple matter to see that in the case of the boundary-value problem (3.8),(3.9) the conditions (3.13) are replaced by (3.6).

3.3. Extension, Bending and Torsion

Let $\mathcal{R}$ be the set of all rigid displacement fields. In view of Corollary 3.1 we are led to seek a solution $\underline{u}^0$ of the problem (P_1) such that $\underline{u}^0_{,3} \in \mathcal{R}$.

Theorem 3.3. Let J be the set of all vector fields $\underline{u} \in C^1(\bar{B}) \cap C^2(B)$ such that $\underline{u}_{,3} \in \mathcal{R}$. Then there exists a vector field $\underline{u}^0 \in J$ which is solution of the problem (P_1).

Proof. Let $\underline{u}^0 \in C^1(\bar{B}) \cap C^2(B)$ such that

$$\underline{u}^0_{,3} = \underline{\alpha} + \underline{\beta} \times \underline{x} ,$$

where $\underline{\alpha}$ and $\underline{\beta}$ are constant vectors. Then it follows that

$$\begin{aligned} u^0_\alpha &= -\frac{1}{2} a_\alpha x_3^2 - a_4 \varepsilon_{\alpha\beta} x_\beta x_3 + w_\alpha , \\ u^0_3 &= (a_\rho x_\rho + a_3) x_3 + w_3 , \end{aligned} \tag{3.14}$$

except for an additive rigid displacement field. Here $\underline{w}$ is an arbitrary vector field independent of x_3 such that $\underline{w} \in C^1(\bar{\Sigma}) \cap C^2(\Sigma)$, and we have used the notations $a_\alpha = \varepsilon_{\rho\alpha} \beta_\rho$, $a_3 = \alpha_3$, $a_4 = \beta_3$. It follows from (3.14) that

$$u^0_{k,\alpha} = a_\alpha x_3 \delta_{k3} - a_4 \varepsilon_{\beta\alpha} x_3 \delta_{k\beta} + w_{k,\alpha} ,$$

$$u^0_{k,3} = (a_\rho x_\rho + a_3) \delta_{k3} - \delta_{k\alpha} a_\alpha x_3 - \delta_{k\alpha} a_4 \varepsilon_{\alpha\beta} x_\beta .$$

The stress-displacement relations imply that

$$S_{ij}(\underline{u}^0) = C_{ij33}(a_\rho x_\rho + a_3) - a_4 C_{ij\alpha 3} \varepsilon_{\alpha\beta} x_\beta + T_{ij}(\underline{w}), \tag{3.15}$$

where

$$T_{ij}(\underline{w}) = C_{ijk\alpha} w_{k,\alpha} , \tag{3.16}$$

Clearly, $T_{ij}(\underline{w})$ are independent of the axial coordinate.

The equations of equilibrium and the conditions on the lateral boundary reduce to

$$(T_{i\alpha}(\underline{w}))_{,\alpha} + g_i = 0 \text{ on } \Sigma, \quad T_{i\alpha}(\underline{w})n_\alpha = q_i \text{ on } \Gamma , \tag{3.17}$$

where

$$\begin{aligned} g_i &= a_\rho(C_{i\alpha 33}x_\rho)_{,\alpha} + a_3 C_{i\alpha 33,\alpha} - a_4 \varepsilon_{\rho\beta}(C_{i\alpha\rho 3}x_\beta)_{,\alpha} , \\ q_i &= (a_4 \varepsilon_{\rho\beta} C_{i\alpha\rho 3}x_\beta - a_\rho C_{i\alpha 33}x_\rho - a_3\, C_{i\alpha 33})n_\alpha . \end{aligned} \tag{3.18}$$

The relations (3.16)-(3.18) constitute a generalized plane strain problem. It follows from (3.18) and the divergence theorem that the necessary and sufficient conditions to solve this problem are satisfied for any constants a_1, a_2, a_3 and a_4. We denote by $\underline{w}^{(j)}$ a solution of the boundary-value problem (3.16)-(3.18) when $a_i = \delta_{ij}$, $a_4 = 0$, and by $\underline{w}^{(4)}$ a solution of the boundary-value problem (3.16)-(3.18) corresponding to $a_i = 0$, $a_4 = 1$. Clearly, we have

$$\underline{w} = \sum_{i=1}^{4} a_i\, \underline{w}^{(i)} . \tag{3.19}$$

Thus, $\underline{w}^{(s)}$ are characterized by the equations

$$\begin{aligned} (T_{i\alpha}(\underline{w}^{(\beta)}))_{,\alpha} + (C_{i\alpha 33}x_\beta)_{,\alpha} &= 0 \ (\beta = 1,2), \\ (T_{i\alpha}(\underline{w}^{(3)}))_{,\alpha} + C_{i\alpha 33,\alpha} &= 0, \\ (T_{i\alpha}(\underline{w}^{(4)}))_{,\alpha} - \varepsilon_{\rho\beta}(C_{i\alpha\rho 3}x_\beta)_{,\alpha} &= 0 \text{ on } \Sigma, \end{aligned} \tag{3.20}$$

and the boundary conditions

$$\begin{aligned} T_{i\alpha}(\underline{w}^{(\beta)})n_\alpha = -C_{i\alpha 33}x_\beta n_\alpha , \quad & T_{i\alpha}(\underline{w}^{(3)})n_\alpha = -C_{i\alpha 33}n_\alpha, \\ T_{i\alpha}(\underline{w}^{(4)})n_\alpha = \varepsilon_{\rho\beta} C_{i\alpha\rho 3}x_\beta n_\alpha & \text{ on } \Gamma . \end{aligned} \tag{3.21}$$

In what follows we assume that the displacement fields $\underline{w}^{(s)}$ (s=1,2,3,4) are known. The vector field $\underline{u}^o$ can be written in the form

$$\underline{u}^o = \sum_{j=1}^{4} a_j\, \underline{u}^{(j)} , \tag{3.22}$$

where $\underline{u}^{(j)}$ are defined by

$$u_\alpha^{(\beta)} = -\frac{1}{2} x_3^2 \delta_{\alpha\beta} + w_\alpha^{(\beta)} , \quad u_3^{(\beta)} = x_\beta x_3 + w_3^{(\beta)} \quad (\beta=1,2) ,$$
$$u_\alpha^{(3)} = w_\alpha^{(3)} , \quad u_3^{(3)} = x_3 + w_3^{(3)}, \quad u_\alpha^{(4)} = \varepsilon_{\beta\alpha} x_\beta x_3 + w_\alpha^{(4)} , \tag{3.23}$$
$$u_3^{(4)} = w_3^{(4)} .$$

It follows from (3.15) and (3.22) that

$$\underline{S}(\underline{u}^o) = \sum_{j=1}^{4} a_j \, \underline{S}(\underline{u}^{(j)}), \tag{3.24}$$

where

$$S_{ij}(\underline{u}^{(\alpha)}) = C_{ij33} x_\alpha + T_{ij}(\underline{w}^{(\alpha)}),$$
$$S_{ij}(\underline{u}^{(3)}) = C_{ij33} + T_{ij}(\underline{w}^{(3)}), \quad S_{ij}(\underline{u}^{(4)}) = -C_{ij\alpha 3} \varepsilon_{\alpha\beta} x_\beta + \tag{3.25}$$
$$+ T_{ij}(\underline{w}^{(4)}).$$

In view of (3.20) and (3.21),

$$\operatorname{div} \underline{S}(\underline{u}^{(j)}) = \underline{0} \text{ on } B , \quad \underline{s}(\underline{u}^{(j)}) = \underline{0} \text{ on } \Pi \quad (j=1,2,3,4), \tag{3.26}$$

so that $\underline{u}^{(j)} \in \mathcal{D}^*$ $(j=1,2,3,4)$.

The conditions on the terminal cross-section Σ_1 are

$$R_\alpha(\underline{u}^o) = 0, \quad R_3(\underline{u}^o) = F_3, \quad \underline{H}(\underline{u}^o) = \underline{M}. \tag{3.27}$$

By Theorem 3.1 and since $\underline{u}^o_{,3} \in \mathcal{R}$, we obtain

$$R_\alpha(\underline{u}^o) = \varepsilon_{\beta\alpha} H_\alpha(\underline{u}^o_{,3}) = 0,$$

so that the first two conditions from (3.27) are satisfied. The remaining conditions furnish the following system for the constants a_s $(s=1,2,3,4)$

$$\sum_{i=1}^{4} D_{\alpha i} a_i = \varepsilon_{\alpha\beta} M_\beta , \quad \sum_{i=1}^{4} D_{3i} a_i = -F_3, \quad \sum_{i=1}^{4} D_{4i} a_i = -M_3, \tag{3.28}$$

where

$$D_{\alpha i} = \int_\Sigma x_\alpha S_{33}(\underline{u}^{(i)}) da, \quad D_{3i} = \int_\Sigma S_{33}(\underline{u}^{(i)}) da ,$$
$$D_{4i} = \int_\Sigma \varepsilon_{\alpha\beta} x_\alpha S_{3\beta}(\underline{u}^{(i)}) da \qquad (i=1,2,3,4) . \tag{3.29}$$

Clearly, the constants D_{rs} (r,s=1,2,3,4) can be calculated after the displacement fields $\underline{w}^{(i)}$(i=1,2,3,4) are determined. Let us prove that the system (3.28) can always be solved for a_1, a_2, a_3 and a_4.

In view of (1.5) and (3.22),

$$U(\underline{u}^0) = \frac{1}{2} \sum_{i,j=1}^{4} \langle \underline{u}^{(i)}, \underline{u}^{(j)} \rangle \, a_i a_j \, .$$

Since $\underline{C}$ is positive definite and $\underline{u}^{(i)}$ is not a rigid displacement, we find that

$$\det \langle \underline{u}^{(i)}, \underline{u}^{(j)} \rangle \neq 0 \, . \tag{3.30}$$

Recall that $\underline{u}^{(i)} \in \mathfrak{D}^*$ (i=1,2,3,4). It follows from (3.23),(3.25),(1.8) and (1.9) that

$$\langle \underline{u}^{(i)}, \underline{u}^{(\alpha)} \rangle = \frac{1}{2} h^2 R_\alpha(\underline{u}^{(i)}) + h \, D_{\alpha i} \, ,$$

$$\langle \underline{u}^{(i)}, \underline{u}^{(3)} \rangle = h \, D_{3i}, \quad \langle \underline{u}^{(i)}, \underline{u}^{(4)} \rangle = hD_{4i}$$

$$(i=1,2,3,4).$$

Since $\underline{u}^{(i)} \in \mathfrak{D}^*$ and $\underline{u}^{(i)}_{,3} \in \mathcal{R}$, by Theorem 3.1 we have $R_\alpha(\underline{u}^{(i)}) = 0$ (i=1,2,3,4). Thus, we obtain

$$\langle \underline{u}^{(i)}, \underline{u}^{(j)} \rangle = h \, D_{ji} . \tag{3.31}$$

It follows from (3.30) and (3.31) that

$$\det(D_{rs}) \neq 0 \, , \tag{3.32}$$

so that the system (3.28) uniquely determines the constants a_i (i = = 1,2,3,4). Thus, we have proved that the constants a_s (s=1,2,3,4) and the vector field $\underline{w}$ can be determined so that $\underline{u}^0 \in K_I(F_3, M_1, M_2, M_3)$.□

<u>Remark.</u> The proof of Theorem 3.3 offers a constructive procedure to obtain a solution of the problem (P_1) for anisotropic elastic bodies. This solution is given by (3.22) and (3.23) where the vector fields $\underline{w}^{(j)}$ (j=1,2,3,4) are characterized by the boundary-value problems (3.20),(3.21), and the constants a_s (s=1,2,3,4) are determined by (3.28).

When the body is isotropic, (3.20) and (3.21) imply

$$w_3^{(i)} = 0 \ (i=1,2,3) \, , \quad w_\alpha^{(4)} = 0 \, , \quad w_3^{(4)} = \varphi \, ,$$

where the function φ is characterized by

$$(\mu\varphi_{,\alpha})_{,\alpha} = \varepsilon_{\rho\beta}(\mu x_\beta)_{,\rho} \text{ on } \Sigma, \quad \varphi_{,\alpha} n_\alpha = \varepsilon_{\rho\beta} x_\beta n_\rho \text{ on } \Gamma .$$

Moreover, the functions $w_\alpha^{(r)}$ $(r=1,2,3)$ are the components of the displacement fields in the following plane strain problems

$$T_{\alpha\beta}(\underline{w}^{(j)}) = \lambda w^{(j)}_{\rho,\rho}\delta_{\alpha\beta} + \mu(w^{(j)}_{\alpha,\beta} + w^{(j)}_{\beta,\alpha}) \qquad (j = 1,2,3),$$

$$(T_{\alpha\beta}(\underline{w}^{(\rho)}))_{,\beta} + (\lambda x_\rho)_{,\alpha} = 0 \qquad (\rho = 1,2),$$

$$(T_{\alpha\beta}(\underline{w}^{(3)}))_{,\beta} + \lambda_{,\alpha} = 0 \qquad \text{on } \Sigma ,$$

$$T_{\alpha\beta}(\underline{w}^{(\rho)})n_\beta = -\lambda x_\rho n_\alpha , \quad T_{\alpha\beta}(\underline{w}^{(3)})n_\beta = -\lambda n_\alpha \quad \text{on } \Gamma .$$

Assume now that the body is homogeneous and isotropic. In order to find the displacement field $\underline{w}^{(1)}$ we note that the corresponding equilibrium equations and boundary conditions are satisfied if one chooses $T_{\alpha\beta}(\underline{w}^{(1)}) = -\lambda x_1 \delta_{\alpha\beta}$. Clearly, since λ is constant these stresses satisfy the compatibility condition. It follows from the constitutive equations that

$$w^{(1)}_{1,1} = w^{(1)}_{2,2} = -\frac{\lambda}{2(\lambda+\mu)} x_1 , \quad w^{(1)}_{1,2} + w^{(1)}_{2,1} = 0.$$

The integration of these equations yields

$$w_1^{(1)} = -\frac{\lambda}{4(\lambda+\mu)}(x_1^2 - x_2^2), \quad w_2^{(1)} = -\frac{\lambda}{2(\lambda+\mu)} x_1 x_2 ,$$

modulo a plane rigid displacement. In a similar way we obtain

$$w_1^{(2)} = -\frac{\lambda}{2(\lambda+\mu)} x_1 x_2 , \quad w_2^{(2)} = \frac{\lambda}{4(\lambda+\mu)}(x_1^2 - x_2^2) ,$$

$$w_\alpha^{(3)} = -\frac{\lambda}{2(\lambda+\mu)} x_\alpha .$$

In this case the function φ is characterized by (1.26) and the constants D_{rs} $(r,s=1,2,3,4)$ reduce to

$$D_{\alpha\beta} = E\, I_{\alpha\beta}, \quad D_{\alpha 3} = E\, A\, x_\alpha^0 , \quad D_{33} = E\, A , \quad D_{i4} = 0, \quad D_{44} = \mu D,$$

where we have used the notations from (1.30). It is a simple matter to see that we rediscover Saint-Venant's solution.

3.4. Application

In [20], Chiriţă applied the above result in order to obtain the solution of the problem (P_1) for a homogeneous and anisotropic circular cylinder. In this section we present this solution.

We assume that Γ is a circle of radius a. Suppose that the x_3-axis passes through the center of the cross-section. First we determine the vector fields $\underline{w}^{(s)}$ (s=1,2,3,4). We try to obtain the vector field $\underline{w}^{(4)}$ assuming that

$$T_{ij}(\underline{w}^{(4)}) = A_{ij\alpha}x_\alpha ,$$

where $A_{ij\alpha}$ are unknown constants. The corresponding equilibrium equations and boundary conditions imply that

$$A_{ij\alpha} = \varepsilon_{\beta\alpha}[C_{ij\beta 3} + \delta_{i3}\delta_{j3}B_\beta + (\delta_{i3}\delta_{j\beta} + \delta_{i\beta}\delta_{j3})B_3],$$

where B_i are arbitrary constants. If the elasticity tensor is invertible, then its inverse $\underline{K} = \underline{C}^{-1}$ defines the relation

$$\underline{E}(\underline{u}) = \underline{K}[\underline{S}(\underline{u})].$$

The equation $E_{33}(\underline{w}^{(4)}) = 0$ implies that

$$B_\alpha = -2\, K^{-1}_{3333}\, K_{\alpha 333}\, B_3 .$$

The compatibility equations are satisfied if and only if

$$B_3 = -\frac{1}{2} Q ,$$

where

$$Q = (A_{\alpha 3\alpha 3})^{-1} , \quad A_{ijrs} = K_{ijrs} - K^{-1}_{3333}\, K_{ij33}\, K_{rs33} .$$

Thus, we conclude that

$$T_{ij}(\underline{w}^{(4)}) = \varepsilon_{\alpha\beta}x_\beta\{C_{ij\alpha 3} + [\delta_{i3}\delta_{j3}\, K^{-1}_{3333}\, K_{\alpha 333} - \frac{1}{2}(\delta_{i3}\delta_{j\alpha} + \delta_{i\alpha}\delta_{j3})]Q\}.$$

The displacement field $\underline{w}^{(4)}$ is given by

$$w_\alpha^{(4)} = \varepsilon_{\beta\alpha}(A_{\eta\rho\eta 3}\, x_\beta x_\rho - \frac{1}{2} A_{\eta\rho\beta 3}\, x_\eta x_\rho)Q ,$$

$$w_3^{(4)} = Q\, A_{\rho 3\alpha 3}\, \varepsilon_{\beta\alpha} x_\beta x_\rho .$$

In a similar manner, we arrive at

$$T_{ij}(\underline{w}^{(\alpha)}) = -C_{ij33}x_\alpha + \delta_{i3}\delta_{j3}K_{3333}^{-1}x_\alpha + \varepsilon_{\alpha\beta}K_{\beta 333}K_{3333}^{-1}(\varepsilon_{\eta\rho}C_{ij\eta 3}x_\rho - T_{ij}(\underline{w}^{(4)})),$$

$$T_{ij}(\underline{w}^{(3)}) = -C_{ij33} + \delta_{i3}\delta_{j3}\, K_{3333}^{-1} ,$$

$$w_i^{(\alpha)} = K_{3333}^{-1}(K_{i\rho 33}x_\rho x_\alpha - \frac{1}{2} K_{\eta\rho 33}\, x_\eta x_\rho \delta_{i\alpha} - \varepsilon_{\alpha\eta}K_{\eta 333}w_i^{(4)}) ,$$

$$w_i^{(3)} = K_{3333}^{-1}\, (K_{i\rho 33}\, x_\rho + \delta_{i3}\, K_{3\rho 33}\, x_\rho).$$

In view of (3.22) and (3.23), we conclude that the solution of the problem (P_1) has the form

$$\begin{aligned} u_\alpha^o = {} & - \frac{1}{2}\, a_\alpha x_3^2 - a_4\varepsilon_{\alpha\beta}x_\beta x_3 + K_{3333}^{-1}\, K_{\alpha\rho 33}\, x_\rho(a_\gamma x_\gamma + a_3) + \\ & + \varepsilon_{\lambda\alpha}(A_{\eta\rho\eta 3}\, x_\lambda - \frac{1}{2}\, A_{\eta\rho\lambda 3}\, x_\eta)x_\rho(a_4 + \varepsilon_{\gamma\beta}a_\beta K_{\gamma 333}\, K_{3333}^{-1})Q - \\ & - \frac{1}{2}\, K_{3333}^{-1}\, K_{\gamma\rho 33}\, a_\alpha x_\gamma x_\rho , \end{aligned}$$

$$\begin{aligned} u_3^o = {} & (a_\gamma x_\gamma + a_3)x_3 + K_{3333}^{-1}\, K_{3\rho 33}\, a_\alpha x_\alpha x_\rho + 2a_3 x_\alpha K_{\alpha 333}\, K_{3333}^{-1} + \\ & + \varepsilon_{\alpha\beta}x_\alpha x_\rho A_{\rho 3\beta 3}\, Q\, (a_4 + \varepsilon_{\eta\gamma}K_{\eta 333}\, K_{3333}^{-1}\, a_\gamma). \end{aligned}$$

Further,

$$S_{3\alpha}(\underline{u}^o) = - \frac{1}{2}\varepsilon_{\alpha\beta}x_\beta(a_4 + \varepsilon_{\gamma\rho}a_\rho K_{\gamma 333}\, K_{3333}^{-1})Q ,$$

$$\begin{aligned} S_{33}(\underline{u}^o) = {} & (a_\alpha x_\alpha + a_3)K_{3333}^{-1} + \\ & + \varepsilon_{\eta\rho}x_\rho K_{3333}^{-1}\, Q\, K_{\eta 333}(a_4 + \varepsilon_{\lambda\beta}a_\beta K_{\lambda 333}\, K_{3333}^{-1}). \end{aligned}$$

A simple calculation shows that the solution of the system (3.28) is given by

$$a_\alpha = \frac{4}{\pi a^4}\varepsilon_{\alpha\beta}(K_{3333}\, M_\beta - K_{\beta 333}\, M_3),$$

$$a_3 = -\frac{1}{\pi a^2}K_{3333}F_3\ ,\quad a_4 = -\frac{4}{\pi a^4}(K_{\nu 3\nu 3}M_3 - K_{\alpha 333}M_\alpha).$$

3.5. Flexure

The flexure problem consists in finding an equilibrium displacement field $\underline{u}$ that satisfies the conditions

$$\underline{s}(\underline{u}) = \underline{0} \text{ on } \Pi,\ R_\alpha(\underline{u}) = F_\alpha\ ,\ R_3(\underline{u}) \doteq 0\ ,\quad \underline{H}(\underline{u}) = \underline{0}\ .$$

Let $\hat{a}$ be the four-dimensional vector (a_1,a_2,a_3,a_4). We shall write $\underline{u}^0\{\hat{a}\}$ for the displacement vector $\underline{u}^0$ defined by (3.22), indicating thus its dependence on the constants a_1,a_2,a_3 and a_4. In view of Corollaries 3.1, 3.2 and Theorem 3.3 it is natural to seek a solution of the flexure problem in the form

$$\underline{u} = \int_0^{x_3}\underline{u}^0\{\hat{b}\}dx_3 + \underline{u}^0\{\hat{c}\} + \underline{w}'\ , \tag{3.33}$$

where $\hat{b} = (b_1,b_2,b_3,b_4)$ and $\hat{c} = (c_1,c_2,c_3,c_4)$ are two constant four-dimensional vectors, and $\underline{w}'$ is a vector field independent of x_3 such that $\underline{w}' \in C^1(\bar{\Sigma})\cap C^2(\Sigma)$.

Theorem 3.4. Let Y be the set of all vector fields of the form (3.33). Then there exists a vector field $\underline{u}' \in Y$ which is solution of the flexure problem.

Proof. Let us prove that the vector field $\underline{w}'$ and the constants b_i, c_i $(i=1,2,3,4)$ can be determined so that $\underline{u}' \in K_{II}(F_1,F_2)$. First, we determine the vector $\hat{b}$. If $\underline{u}' \in K_{II}(F_1,F_2)$, then by Corollary 3.2 and (3.33),

$$\underline{u}^0\{\hat{b}\} \in K_I(0,F_2,-F_1,0). \tag{3.34}$$

In view of (3.28) and (3.34) we find that

$$\sum_{i=1}^{4} D_{\alpha i}\, b_i = -F_\alpha\ , \tag{3.35}$$

$$\sum_{i=1}^{4} D_{3i}\, b_i = 0 \,, \quad \sum_{i=1}^{4} D_{4i}\, b_i = 0.$$

The system (3.35) can always be solved for b_1, b_2, b_3 and b_4.

It follows from (3.22),(3.23) and (3.33) that

$$\begin{aligned} u'_\alpha &= -\frac{1}{6} b_\alpha x_3^3 - \frac{1}{2} c_\alpha x_3^2 - \frac{1}{2} b_4 \varepsilon_{\alpha\beta} x_\beta x_3^2 - c_4 \varepsilon_{\alpha\beta} x_\beta x_3 + \\ &+ \sum_{j=1}^{4} (c_j + x_3 b_j) w_\alpha^{(j)} + w'_\alpha \,, \\ u'_3 &= \frac{1}{2} (b_\rho x_\rho + b_3) x_3^2 + (c_\rho x_\rho + c_3) x_3 + \\ &+ \sum_{j=1}^{4} (c_j + x_3 b_j) w_3^{(j)} + w'_3 \,. \end{aligned} \tag{3.36}$$

In view of (1.1) and (3.25) we obtain

$$\underline{S}(\underline{u}') = \sum_{i=1}^{4} (c_i + x_3 b_i) \underline{S}(\underline{u}^{(i)}) + \underline{k} + \underline{T}(\underline{w}'), \tag{3.37}$$

where

$$k_{ij} = \sum_{r=1}^{4} C_{ijk3} b_r w_k^{(r)} \,.$$

Substituting (3.37) into equations of equilibrium, we find, with the aid of (3.26), that

$$(T_{i\alpha}(\underline{w}'))_{,\alpha} + f'_i = 0 \quad \text{on } \Sigma \,, \tag{3.38}$$

where

$$f'_i = k_{i\alpha,\alpha} + \sum_{j=1}^{4} b_j S_{i3}(\underline{u}^{(j)}).$$

In view of (3.26), the conditions on the lateral boundary reduce to

$$T_{i\alpha}(\underline{w}') n_\alpha = p'_i \quad \text{on } \Gamma \,, \tag{3.39}$$

where $p'_i = -k_{i\alpha} n_\alpha$. The relations (3.38) and (3.39) constitute a generalized plane strain problem. The necessary and sufficient conditions for the existence of a solution of this problem are

$$\int_\Sigma f'_i \, da + \int_\Gamma p'_i \, ds = 0, \quad \int_\Sigma \varepsilon_{\alpha\beta} x_\alpha f'_\beta \, da + \int_\Gamma \varepsilon_{\alpha\beta} x_\alpha p'_\beta \, ds = 0 \,.$$

It is a simple matter to verify that these conditions become

$$\sum_{j=1}^{4} b_j \int_{\Sigma} S_{i3}(\underline{u}^{(j)})da = 0, \quad \sum_{j=1}^{4} b_j \int_{\Sigma} \varepsilon_{\alpha\beta} x_\alpha S_{\beta 3}(\underline{u}^{(j)})da = 0 \; . \tag{3.40}$$

It follows from (3.35) and $R_\alpha(\underline{u}^{(j)}) = 0$ $(j=1,2,3,4)$ that the conditions (3.40) are satisfied. In what follows we assume that the displacement field $\underline{w}'$ is known.

Since $H_\alpha(\underline{u}'_{,3}) = \varepsilon_{\alpha\beta} R_\beta(\underline{u}')$ and $\underline{u}'_{,3} \in K_I(0,F_2,-F_1,0)$ it follows that $R_\alpha(\underline{u}') = F_\alpha$. The conditions $R_3(\underline{u}') = 0$, $\underline{H}(\underline{u}') = \underline{0}$ are satisfied if and only if

$$\sum_{j=1}^{4} D_{ij}\, c_j = A_i \quad (i=1,2,3,4) \; , \tag{3.41}$$

where

$$A_\alpha = - \int_{\Sigma} x_\alpha [k_{33} + T_{33}(\underline{w}')]\, da, \quad A_3 = - \int_{\Sigma} [k_{33} + T_{33}(\underline{w}')]\, da \; ,$$

$$A_4 = - \int_{\Sigma} \varepsilon_{\alpha\beta} x_\alpha [k_{\beta 3} + T_{\beta 3}(\underline{w}')]\, da \; .$$

In view of (3.32), the system (3.41) can always be solved for c_1, c_2, c_3 and c_4. Thus, if $\hat{b}$ and $\hat{c}$ are defined by (3.35) and (3.41), respectively, and the displacement field $\underline{w}'$ is characterized by the generalized plane strain problem (3.38),(3.39), then the displacement field $\underline{u}'$ defined by (3.36) is a solution of the flexure problem. □

We derived the system (3.41) from the conditions $R_3(\underline{u}') = 0$, $\underline{H}(\underline{u}') = \underline{0}$. If we replace these conditions by $R_3(\underline{u}') = F_3$, $\underline{H}(\underline{u}') = \underline{M}$, then we arrive at

$$\sum_{i=1}^{4} D_{\alpha i}\, c_i = \varepsilon_{\alpha\beta} M_\beta + A_\alpha ,$$
$$\sum_{i=1}^{4} D_{3i} c_i = A_3 - F_3 \; , \quad \sum_{i=1}^{4} D_{4i}\, c_i = A_4 - M_3 \; . \tag{3.42}$$

Clearly, if $\hat{b}$ is defined by (3.35), $\hat{c}$ is defined by (3.42), and $\underline{w}'$ is characterized by the boundary-value problem (3.38),(3.39), then $\underline{u}' \in K(\underline{F},\underline{M})$.

In the case of homogeneous and isotropic bodies we rediscover Saint-Venant's solution.

The above method to construct a solution of the flexure problem has been used by Chiriţă [20] to establish the solution for a homogeneous and anisotropic circular cylinder.

3.6. Minimum Energy Characterizations of Solutions

In this section we present minimum strain-energy characterizations of the solutions obtained in Sections 3.3 and 3.5. Analogous theorems for homogeneous and isotropic bodies were given by Sternberg and Knowles [143].

Let Q_I denote the set of all equilibrium displacement fields $\underline{u}$ that satisfy the conditions

$$\begin{gathered} [S_{3i}(\underline{u})](x_1,x_2,0) = [S_{3i}(\underline{u})](x_1,x_2,h) \quad (x_1,x_2)\in\Sigma, \\ R_\alpha(\underline{u}) = 0, \quad R_3(\underline{u}) = F_3, \quad \underline{H}(\underline{u}) = \underline{M}, \\ \underline{s}(\underline{u}) = \underline{0} \quad \text{on } \Pi. \end{gathered} \tag{3.43}$$

<u>Theorem 3.5.</u> Let $\underline{u}^0$ be the solution (3.22) of the problem (P_1) corresponding to the scalar load F_3 and the moment $\underline{M}$. Then

$$U(\underline{u}^0) \leq U(\underline{u}),$$

for every $\underline{u}\in Q_I$, and equality holds only if $\underline{u}=\underline{u}^0$ modulo a rigid displacement.

<u>Proof.</u> Note that $\underline{u}^0\in Q_I$. Let $\underline{u}\in Q_I$ and define

$$\underline{v} = \underline{u} - \underline{u}^0 .$$

Then $\underline{v}$ is an equilibrium displacement field that satisfies

$$\begin{gathered} [S_{3i}(\underline{v})](x_1,x_2,0) = [S_{3i}(\underline{v})](x_1,x_2,h) \quad (x_1,x_2)\in\Sigma, \\ \underline{s}(\underline{v}) = \underline{0} \quad \text{on } \Pi, \quad \underline{R}(\underline{v}) = \underline{0}, \quad \underline{H}(\underline{v}) = \underline{0}. \end{gathered} \tag{3.44}$$

By (1.5) and (1.6),

$$U(\underline{u}) = U(\underline{v}) + U(\underline{u}^0) + \langle \underline{v},\underline{u}^0\rangle .$$

Next, if we apply (1.8) and (1.9) we conclude, with the aid of (3.22) (3.23) and (3.44), that

$$< \underline{v}, \underline{u}^o > = \int_{\Sigma_2} S_{3i}(\underline{v}) u_i^o \, da - \int_{\Sigma_1} S_{3i}(\underline{v}) u_i^o \, da = -\frac{1}{2} h^2 a_\alpha R_\alpha(\underline{v}) +$$

$$+ h[\varepsilon_{\alpha\beta} a_\alpha H_\beta(\underline{v}) - a_3 R_3(\underline{v}) - a_4 H_3(\underline{v})] = 0 .$$

This result implies that

$$U(\underline{u}) = U(\underline{v}) + U(\underline{u}^o) .$$

Thus $U(\underline{u}) \geq U(\underline{u}^o)$, and $U(\underline{u}) = U(\underline{u}^o)$ only if $\underline{v}$ is a rigid displacement.□

Let Q_{II} denote the set of all equilibrium displacement fields $\underline{u}$ that satisfy the conditions

$$\begin{aligned} &\underline{u}_{,3} \in C^1(\bar{B}) \cap C^2(B) , \quad \underline{s}(\underline{u}) = \underline{0} \text{ on } \Pi, \quad R_\alpha(\underline{u}) = F_\alpha , \\ &[S_{3i}(\underline{u}_{,3})](x_1,x_2,0) = [S_{3i}(\underline{u}_{,3})](x_1,x_2,h) \quad (x_1,x_2) \in \Sigma . \end{aligned} \tag{3.45}$$

<u>Theorem 3.6.</u> Let $\underline{u}'$ be the solution (3.36) of the flexure problem corresponding to the loads F_1 and F_2. Then

$$U(\underline{u}'_{,3}) \leq U(\underline{u}_{,3}),$$

for every $\underline{u} \in Q_{II}$, and equality holds only if $\underline{u}_{,3} = \underline{u}'_{,3}$ (modulo a rigid displacement).

<u>Proof.</u> Let $\underline{u} \in Q_{II}$. Since $\underline{u}' \in Q_{II}$ it follows that the field

$$\underline{v} = \underline{u} - \underline{u}' ,$$

is an equilibrium displacement field that satisfies

$$\begin{aligned} &\underline{v}_{,3} \in C^1(\bar{B}) \cap C^2(B), \quad \underline{s}(\underline{v}) = \underline{0} \text{ on } \Pi, \quad R_\alpha(\underline{v}) = 0 , \\ &[S_{3\beta}(\underline{v}_{,3})](x_1,x_2,0) = [S_{3\beta}(\underline{v}_{,3})](x_1,x_2,h) \quad (x_1,x_2) \in \Sigma . \end{aligned} \tag{3.46}$$

According to (1.5), (3.33) and Theorem 3.3, we have

$$U(\underline{u}_{,3}) = U(\underline{v}_{,3} + \underline{u}'_{,3}) = U(\underline{v}_{,3} + \underline{u}^o\{\hat{b}\}) = U(\underline{v}_{,3}) + U(\underline{u}'_{,3}) + < \underline{v}_{,3}, \underline{u}^o\{\hat{b}\} > .$$

By (3.22), (1.8), (1.9) and (3.46),

$$< \underline{v}_{,3}, \underline{u}^o\{\hat{b}\} > = -\frac{1}{2} b_\alpha h^2 R_\alpha(\underline{v}_{,3}) + h[b_1 H_2(\underline{v}_{,3}) - b_2 H_1(\underline{v}_{,3}) - b_3 R_3(\underline{v}_{,3}) -$$

$$- b_4 H_3(\underline{v}_{,3})] .$$

In view of Theorem 3.1 and (3.46), we conclude that $<\underline{v}_{,3},\underline{u}^{0}\{\hat{b}\}> = 0$. Thus,

$$U(\underline{u}_{,3}) = U(\underline{v}_{,3}) + U(\underline{u}'_{,3}) .$$

The desired conclusion is now immediate.

3.7. A Solution of Truesdell's Problem

Truesdell's problem as formulated in Section 1.5 can be set also for anisotropic bodies. Thus we are led to the following problem: to define the functionals $\tau_i\ (\cdot)$ (i=1,2,3,4) on $K_I(F_3,M_1,M_2,M_3)$ such that

$$\sum_{j=1}^{4} D_{\alpha j}\,\tau_j(\underline{u}) = \varepsilon_{\alpha\beta}M_\beta ,$$
$$\sum_{j=1}^{4} D_{3j}\,\tau_j(\underline{u}) = -F_3 ,\quad \sum_{j=1}^{4} D_{4j}\,\tau_j(\underline{u}) = -M_3, \tag{3.47}$$

hold for each $\underline{u}\in K_I(F_3,M_1,M_2,M_3)$.

In order to study this problem we first consider the set Q_I of all equilibrium displacement fields $\underline{u}$ that satisfy the conditions (3.43). Clearly, if $\underline{u}\in Q_I$ then $\underline{u}\in K_I(F_3,M_1,M_2,M_3)$. In view of Theorem 3.5 we are led to consider the real function f of the variables ξ_1, ξ_2, ξ_3 and ξ_4 defined by

$$f = \|\underline{u} - \sum_{j=1}^{4} \xi_j\,\underline{u}^{(j)}\|_e^2 ,$$

where $\underline{u}\in Q_I$ and $\underline{u}^{(j)}$ (j=1,2,3,4) are given by (3.23). By (3.31),

$$f = h \sum_{i,j=1}^{4} D_{ij}\,\xi_i\,\xi_j - 2\sum_{i=1}^{4} \xi_i<\underline{u},\underline{u}^{(i)}> + \|\underline{u}\|_e^2 .$$

Since the matrix (D_{ij}) (i,j=1,2,3,4) is positive definite, f will be a minimum at $(\alpha_1(\underline{u}),\alpha_2(\underline{u}),\alpha_3(\underline{u}),\alpha_4(\underline{u}))$ if and only if $(\alpha_1(\underline{u}),\alpha_2(\underline{u}),\alpha_3(\underline{u}),\alpha_4(\underline{u}))$ is the solution of the following system of equations

$$h\sum_{j=1}^{4} D_{ij}\,\alpha_j(\underline{u}) = <\underline{u},\underline{u}^{(i)}> \quad (i=1,2,3,4) . \tag{3.48}$$

By (3.23),(1.8),(1.9) and (3.43), we obtain

$$<\underline{u},\underline{u}^{(1)}> = \int_{\partial B} \underline{s}(\underline{u})\cdot\underline{u}^{(1)}da = h\int_{\Sigma_2} x_1S_{33}(\underline{u})da - \frac{1}{2}h^2\int_{\Sigma_2} S_{31}(\underline{u})da =$$

$$= -\frac{1}{2}h^2 R_1(\underline{u}) + h H_2(\underline{u}) = h H_2(\underline{u}).$$

In a similar manner we arrive at

$$\langle \underline{u},\underline{u}^{(\alpha)}\rangle = h\,\varepsilon_{\alpha\beta}H_\beta(u), \quad \langle \underline{u},\underline{u}^{(3)}\rangle = -R_3(\underline{u}),$$
$$\langle \underline{u},\underline{u}^{(4)}\rangle = -H_3(\underline{u}). \tag{3.49}$$

It follows from (3.47), (3.48) and (3.49) that $\tau_i(\underline{u}) = \alpha_i(\underline{u})$ (i= =1,2,3,4), for each $\underline{u} \in Q_I$.

On the other hand, by (1.9) we find

$$\langle \underline{u},\underline{u}^{(r)}\rangle = A_r(\underline{u}) \quad (r=1,2,3,4), \tag{3.50}$$

where

$$A_r(\underline{u}) = \int_{\Sigma_2} S_{3i}(\underline{u}^{(r)})u_i\,da - \int_{\Sigma_1} S_{3i}(\underline{u}^{(r)})u_i\,da . \tag{3.51}$$

It follows from (3.48) and (3.50) that

$$\sum_{j=1}^{4} D_{ij}\tau_j(\underline{u}) = \frac{1}{2}A_i(\underline{u}) \quad (i=1,2,3,4). \tag{3.52}$$

The system (3.52) defines $\tau_i(\underline{u})$ (i=1,2,3,4) for every displacement field $\underline{u} \in Q_I$.

Truesdell's problem can be set also for the flexure of anisotropic cylinders: to define the functionals $\gamma_i(\cdot)$ (i=1,2,3,4) on $K_{II}(F_1,F_2)$ such that

$$\sum_{i=1}^{4} D_{\alpha i}\gamma_i(\underline{u}) = -F_i ,$$
$$\sum_{i=1}^{4} D_{3i}\gamma_i(\underline{u}) = 0 , \quad \sum_{i=1}^{4} D_{4i}\gamma_i(\underline{u}) = 0 , \tag{3.53}$$

hold for each $\underline{u} \in K_{II}(F_1,F_2)$.

Let $\mathcal{H}$ denote the set of all equilibrium displacement fields $\underline{u}$ that satisfy the conditions

$$\underline{u}_{,3} \in C^1(\bar{B}) \cap C^2(B), \quad \underline{s}(\underline{u}) = \underline{0} \text{ on } \Pi, \quad R_\alpha(\underline{u}) = F_\alpha,$$
$$R_3(\underline{u}) = 0 , \quad \underline{H}(\underline{u}) = \underline{0} , \tag{3.54}$$
$$[S_{3i}(\underline{u}_{,3})](x_1,x_2,0) = [S_{3i}(\underline{u}_{,3})](x_1,x_2,h) \quad (x_1,x_2) \in \Sigma .$$

Clearly, if $\underline{u} \in \mathcal{H}$ then $\underline{u} \in K_{II}(F_1, F_2)$. Let g be the real function of the variables ζ_i (i=1,2,3,4) defined by

$$g = \| \underline{u}_{,3} - \sum_{i=1}^{4} \zeta_i \, \underline{u}^{(i)} \|_e^2 ,$$

where $\underline{u} \in \mathcal{H}$ and $\underline{u}^{(i)}$ (i=1,2,3,4) are given by (4.10). Clearly, g will be a minimum at $(\beta_1(\underline{u}), \beta_2(\underline{u}), \beta_3(\underline{u}), \beta_4(\underline{u}))$ if and only if $(\beta_1(\underline{u}), \beta_2(\underline{u}), \beta_3(\underline{u}), \beta_4(\underline{u}))$ is the solution of the following system of equations

$$h \sum_{j=1}^{4} D_{ij} \beta_j(\underline{u}) = < \underline{u}_{,3}, \underline{u}^{(i)} > \quad (i=1,2,3,4). \qquad (3.55)$$

Let us prove that $\beta_i(\underline{u}) = \gamma_i(\underline{u})$ (i=1,2,3,4) for every $\underline{u} \in \mathcal{H}$. By (3.23),(1.8),(1.9) and (3.54), we obtain

$$< \underline{u}_{,3}, \underline{u}^{(1)} > = \int_{\partial B} \underline{s}(\underline{u}_{,3}) \cdot \underline{u}^{(1)} da = - \frac{1}{2} h^2 R_1(\underline{u}_{,3}) +$$

$$+ h \, H_2(\underline{u}_{,3}) .$$

In view of Theorem 3.1 we arrive at

$$< \underline{u}_{,3}, \underline{u}^{(1)} > = - h \, R_1(\underline{u}).$$

In a similar manner we find that

$$< \underline{u}_{,3}, \underline{u}^{(\alpha)} > = -h \, R_\alpha(\underline{u}), \quad < \underline{u}_{,3}, u^{(2+\alpha)} > = 0 \ (\alpha = 1,2) . \qquad (3.56)$$

It follows from (3.53)-(3.56) that $\gamma_i(\underline{u}) = \beta_i(\underline{u})$ (i=1,2,3,4) for any $\underline{u} \in \mathcal{H}$. On the other hand, by (1.8) we obtain

$$< \underline{u}_{,3}, \underline{u}^{(i)} > = B_i(\underline{u}) \ (i=1,2,3,4), \qquad (3.57)$$

where

$$B_j(\underline{u}) = \int_{\Sigma_2} S_{3i}(\underline{u}^{(j)}) u_{i,3} da - \int_{\Sigma_1} S_{3i}(\underline{u}^{(j)}) u_{i,3} da.$$

Thus, from (3.55) and (3.57) we conclude that

$$\sum_{j=1}^{4} D_{ij} \gamma_j(\underline{u}) = \frac{1}{h} B_i(\underline{u}) \ (i=1,2,3,4) ,$$

for each $\underline{u} \in \mathcal{H}$. This system defines $\gamma_i(\cdot)$ on the subclass $\mathcal{H}$ of solutions to the flexure problem.

3.8. The Problems of Almansi and Michell

We first consider the Almansi-Michell problem. It is a simple matter to verify that Theorem 2.1 also remains valid for anisotropic bodies where the elasticity field is independent of the axial coordinate. As in Section 2.2 we are led to consider the set V of all vector fields of the form

$$\int_0^{x_3}\int_0^{x_3} \underline{u}^0\{\hat{b}\}dx_3dx_3 + \int_0^{x_3} \underline{u}^0\{\hat{c}\}dx_3 + \underline{u}^0\{\hat{d}\} + x_3\,\underline{w}' + \underline{w}'', \tag{3.58}$$

where $\hat{b}$, $\hat{c}$ and $\hat{d}$ are four-dimensional constant vectors, and $\underline{w}'$ and $\underline{w}''$ are vector fields independent of x_3 such that $\underline{w}', \underline{w}'' \in C^1(\bar{\Sigma}) \cap C^2(\Sigma)$. Here $\underline{u}^0\{\hat{a}\}$ is defined by (3.22).

We assume for the remainder of this section that the body force and surface force belong to C^{∞}.

<u>Theorem 3.7.</u> Let B be anisotropic and assume that the elasticity field is independent of the axial coordinate. Then there exists a vector field $\underline{u}'' \in V$ such that $\underline{u}'' \in K_{III}(\underline{F},\underline{M},\underline{f},\underline{p})$.

<u>Proof.</u> If $\underline{u}'' \in V$ and $\underline{u}'' \in K_{III}(\underline{F},\underline{M},\underline{f},\underline{p})$, then by Theorem 3.3, Corollary 2.1 and (3.58),

$$\int_0^{x_3} \underline{u}^0\{\hat{b}\}dx_3 + \underline{u}^0\{\hat{c}\} + \underline{w}' \in K(\underline{G},\underline{Q}) \ ,$$

where $\underline{G}$ and $\underline{Q}$ are defined by (2.3). In view of (3.35) and (3.38) we find that $\hat{b}$ is given by

$$\begin{gathered} \sum_{i=1}^{4} D_{\alpha i}b_i = -\int_{\Sigma} f_{\alpha}da - \int_{\Gamma} p_{\alpha}ds \ , \\ \sum_{i=1}^{4} D_{3i}b_i = 0 \ , \quad \sum_{i=1}^{4} D_{4i}b_i = 0 \ , \end{gathered} \tag{3.59}$$

and $\underline{w}'$ is characterized by

$$\begin{gathered} (T_{i\alpha}(\underline{w}'))_{,\alpha} + \sum_{r=1}^{4} b_r[(C_{i\alpha k3}w_k^{(r)})_{,\alpha} + S_{i3}(\underline{u}^{(r)})] = 0 \text{ on } \Sigma, \\ T_{i\alpha}(\underline{w}')n_{\alpha} = -\sum_{r=1}^{4} C_{i\alpha k3}\, b_r w_k^{(r)}\, n_{\alpha} \quad \text{on } \Gamma . \end{gathered} \tag{3.60}$$

Moreover, in view of (3.42), the vector $\hat{c}$ is determined by

$$\sum_{j=1}^{4} D_{ij}\, c_j = C_i \quad (i=1,2,3,4)\ , \tag{3.61}$$

where

$$C_\alpha = -\int_\Gamma x_\alpha p_3 ds - \int_\Sigma x_\alpha f_3 da - F_\alpha - \int_\Sigma x_\alpha [k'_{33} + T_{33}(w')]\, da\ ,$$

$$C_3 = -\int_\Gamma p_3 ds - \int_\Sigma f_3 da - \int_\Sigma [k'_{33} + T_{33}(w')]\, da\ ,$$

$$C_4 = -\int_\Gamma \varepsilon_{\alpha\beta} x_\alpha p_\beta ds - \int_\Sigma \varepsilon_{\alpha\beta} x_\alpha f_\beta da - \int_\Sigma \varepsilon_{\alpha\beta} x_\alpha [k'_{\beta 3} + T_{\beta 3}(\underline{w}')]\, da,$$

$$k'_{ij} = \sum_{r=1}^{4} C_{ijk3}\, b_r\, w_k^{(r)}.$$

It follows from (3.22),(3.23) and (3.58) that

$$\begin{aligned} u''_\alpha &= -\frac{1}{24} b_\alpha x_3^4 - \frac{1}{6} c_\alpha x_3^3 - \frac{1}{2} d_\alpha x_3^2 - \varepsilon_{\alpha\beta} x_\beta (\frac{1}{6} b_4 x_3^3 + \frac{1}{2} c_4 x_3^2 + \\ &\quad + d_4 x_3) + \sum_{j=1}^{4} (d_j + c_j x_3 + \frac{1}{2} b_j x_3^2) w_\alpha^{(j)} + x_3 w'_\alpha + w''_\alpha\ , \\ u''_3 &= \frac{1}{6} (b_\rho x_\rho + b_3) x_3^3 + \frac{1}{2} (c_\rho x_\rho + c_3) x_3^2 + (d_\rho x_\rho + d_3) x_3 + \\ &\quad + \sum_{j=1}^{4} (d_j + c_j x_3 + \frac{1}{2} b_j x_3^2) w_3^{(j)} + x_3 w'_3 + w''_3\ . \end{aligned} \tag{3.62}$$

By (1.1),(3.25) and (3.62),

$$\begin{aligned} S_{ij}(\underline{u}'') &= \sum_{r=1}^{4} (d_r + c_r x_3 + \frac{1}{2} b_r x_3^2) S_{ij}(\underline{u}^{(r)}) + x_3 k'_{ij} + k''_{ij} + \\ &\quad + T_{ij}(\underline{w}'') + x_3\, T_{ij}(\underline{w}'), \end{aligned} \tag{3.63}$$

where

$$k''_{ij} = C_{ijk3}\, w'_k + \sum_{s=1}^{4} c_s\, C_{ijk3}\, w_k^{(s)},$$

The equations of equilibrium and the condition on the lateral boundary reduce to

$$(T_{i\alpha}(\underline{w}''))_{,\alpha} + h_i = 0 \text{ on } \Sigma ,$$
$$T_{i\alpha}(\underline{w}'')n_\alpha = q_i \quad \text{on } \Gamma , \tag{3.64}$$

where

$$h_i = f_i + k''_{i\alpha,\alpha} + T_{i3}(\underline{w}') + k'_{i3} + \sum_{r=1}^{4} c_r S_{i3}(\underline{u}^{(r)}),$$
$$q_i = p_i - k''_{i\alpha} n_\alpha .$$

By the divergence theorem we find that

$$\int_\Sigma h_i da + \int_\Gamma q_i ds = \int_\Sigma f_i da + \int_\Gamma p_i ds - R_i(\underline{u}''_{,3}) = G_i - R_i(\underline{u}''_{,3}) ,$$

$$\int_\Sigma \varepsilon_{\alpha\beta} x_\alpha h_\beta da + \int_\Gamma \varepsilon_{\alpha\beta} x_\alpha q_\beta ds = \int_\Sigma \varepsilon_{\alpha\beta} x_\alpha f_\beta da + \int_\Gamma \varepsilon_{\alpha\beta} x_\alpha p_\beta da -$$
$$- H_3(\underline{u}''_{,3}) = Q_3 - H_3(\underline{u}''_{,3}) .$$

Note that the conditions

$$R_i(\underline{u}''_{,3}) = G_i \ , \ H_3(\underline{u}''_{,3}) = Q_3 \quad ,$$

were used to obtain (3.59) and (3.61).
Thus we conclude that the necessary and sufficient conditions for the existence of a solution of the boundary-value problem (3.64) are satisfied.

It follows from (3.61) and (3.63) that

$$H_\alpha(\underline{u}''_{,3}) = \varepsilon_{\beta\alpha}\Big(\sum_{i=1}^{4} D_{\beta i} c_i + \int_\Sigma x_\beta[k'_{33} + T_{33}(\underline{w}')] da\Big) =$$
$$= \varepsilon_{\beta\alpha}\Big(\int_\Gamma x_\beta p_3 ds + \int_\Sigma x_\beta f_3 da\Big) + \varepsilon_{\alpha\beta} F_\beta . \tag{3.65}$$

In view of Theorem 2.1,

$$H_\alpha(\underline{u}''_{,3}) = \varepsilon_{\alpha\beta}\Big(\int_\Gamma x_\beta p_3 ds + \int_\Sigma x_\beta f_3 da\Big) + \varepsilon_{\alpha\beta} R_\beta(\underline{u}'') . \tag{3.66}$$

It follows from (3.65) and (3.66) that $R_\alpha(\underline{u}'') = F_\alpha$. The conditions $R_3(\underline{u}'') = F_3$, $\underline{H}(\underline{u}'') = \underline{M}$ reduce to

$$\sum_{s=1}^{4} D_{rs}\, d_s = E_r \quad (r=1,2,3,4) \quad , \tag{3.67}$$

where

$$E_\alpha = \varepsilon_{\alpha\beta}M_\beta - \int_\Sigma x_\alpha[k''_{33} + T_{33}(\underline{w}'')]da \; ,$$

$$E_3 = - F_3 - \int_\Sigma [k''_{33} + T_{33}(\underline{w}'')]da,$$

$$E_4 = - M_3 - \int_\Sigma \varepsilon_{\alpha\beta}x_\alpha[k''_{\beta 3} + T_{\beta 3}(\underline{w}'')]da \; .$$

The system (3.67) determines the vector $\hat{d}$. Thus we determined the vector $\hat{b}, \hat{c}$ and $\hat{d}$, and the vector fields $\underline{w}'$ and $\underline{w}''$ such that $\underline{u}'' \in K_{III}(\underline{F},\underline{M}, \underline{f},\underline{p})$. □

We consider now the Almansi problem. Let $\underline{u}^*$ be an equilibrium displacement field on B corresponding to the body force $\underline{f} = \underline{g}\, x_3^n$, that satisfies the conditions

$$\underline{s}(\underline{u}^*) = \underline{q}\, x_3^n \text{ on } \Pi, \quad \underline{R}(\underline{u}^*) = \underline{0} \; , \quad \underline{H}(\underline{u}^*) = \underline{0} \; , \tag{3.68}$$

where $\underline{g}$ and $\underline{p}$ are vector fields independent of x_3, and n is a positive integer or zero.

Let $\underline{u}$ be an equilibrium displacement field on B corresponding to the body force $\underline{f} = \underline{g}\, x^{n+1}$, that satisfies the conditions

$$\underline{s}(\underline{u}) = \underline{q}\, x_3^{n+1} \text{ on } \Pi \; , \; \underline{R}(\underline{u}) = \underline{0} \; , \quad \underline{H}(\underline{u}) = \underline{0} \; . \tag{3.69}$$

In view of the results obtained in Section 2.3, the Almansi problem reduces to the problem of finding a vector field $\underline{u}$ once the vector field $\underline{u}^*$ is known. As in Section 2.3 we are led to seek the vector field $\underline{u}$ in the form

$$\underline{u} = (n+1)\left[\int_0^{x_3} \underline{u}^* dx_3 + \underline{u}^0\{\hat{a}\} + \underline{w}\,\right], \tag{3.70}$$

where $\hat{a} = (a_1,a_2,a_3,a_4)$ is an unknown vector, $\underline{u}^0\{\hat{a}\}$ is given by (3.22) and $\underline{w}$ is an unknown vector field independent of x_3.

By (3.25) and (3.70),

$$S_{ij}(\underline{u}) = (n+1)\left[\int_0^{x_3} S_{ij}(\underline{u}^*)dx_3 + \sum_{r=1}^{4} a_r S_{ij}(\underline{u}^{(r)}) + T_{ij}(\underline{w}) + g_{ij}\right],$$

where

$$g_{ij} = C_{ijk3}\, u_k^*(x_1,x_2,0) .$$

The equilibrium equations and the conditions on the lateral boundary reduce to

$$(T_{i\alpha}(\underline{w}))_{,\alpha} + h_i' = 0 \text{ on } \Sigma,$$
$$T_{i\alpha}(\underline{w})n_\alpha = q_i' \text{ on } \Gamma, \tag{3.71}$$

where

$$h_i' = g_{i\alpha,\alpha} + (S_{i3}(\underline{u}^*))(x_1,x_2,0), \quad q_i' = - g_{i\alpha}n_\alpha.$$

Clearly, we have

$$\int_\Sigma h_i' da + \int_\Gamma q_i' ds = - R_i(\underline{u}^*) = 0,$$

$$\int_\Sigma \varepsilon_{\alpha\beta} x_\alpha h_\beta' da + \int_\Gamma \varepsilon_{\alpha\beta} x_\alpha q_\beta' ds = - Q_3(\underline{u}^*) = 0 .$$

Thus, the necessary and sufficient conditions for the existence of a solution to the boundary-value problem (3.71) are satisfied. We conclude that the vector field $\underline{w}$ is characterized by the generalized plane strain problem (3.71).

In view of Theorem 2.1, $R_\alpha(\underline{u}) = \varepsilon_{\beta\alpha} H_\beta[(n+1)\underline{u}^*] = 0$. The conditions $R_3(\underline{u}) = 0$, $\underline{H}(\underline{u}) = \underline{0}$ imply that

$$\sum_{s=1}^{4} D_{rs}\, a_s = k_r \quad (r=1,2,3,4) , \tag{3.72}$$

where

$$k_\alpha = - \int_\Sigma x_\alpha (T_{33}(\underline{w}) + g_{33}) da ,$$

$$k_3 = - \int_\Sigma (T_{33}(\underline{w}) + g_{33}) da ,$$

$$k_4 = - \int_\Sigma \varepsilon_{\alpha\beta} x_\alpha (T_{3\beta}(\underline{w}) + g_{3\beta}) da .$$

Thus, the constant vector $\hat{a}$ is determined by (3.72).

The solution established here coincides with the solution established in [67] by the semi-inverse method. Specific applications are presented in [49].

4. HETEROGENEOUS MEDIA

4.1. Preliminaries

This chapter is concerned with the deformation of a heterogeneous cylinder, composed of two elastic materials with different elasticities.

The equilibrium problem for heterogeneous elastic media was studied in various papers (see, for example, Muskhelishvili [111], Kupradze, Gegelia, Bashelishvili and Burchuladze [92], and Fichera [37]). Fichera [37] was the first to consider the case of anisotropic bodies, with general elasticities. The deformation of heterogeneous cylinders has also been studied extensively in the literature (see, e.g., [111], [139],[12],[73],[48]). Most of the papers dealing with this problem are restricted to piecewise homogeneous cylinders. In [66],[68], by use of the semi-inverse method, we established a solution of the relaxed Saint-Venant problem for a cylinder composed of two different nonhomogeneous elastic materials, where the elastic coefficients are independent of the axial coordinate.

In the first part of this chapter we present in a rational form the results concerning the relaxed Saint-Venant's problem for heterogeneous cylinders. Then, the problems of Almansi and Michell are studied. Applications to the problem of thermoelastic deformation of composed cylinders are presented.

Let Σ be a C^1-smooth domain (cf. [37], p.369) and let Γ_1 and Γ_2 be complementary subsets of Γ. Let Γ_0 be a curve contained in Σ with the property that $\overline{\Gamma}_0 \cup \overline{\Gamma}_\rho$ ($\rho = 1,2$) is the boundary of a regular domain A_ρ contained in Σ such that $A_1 \cap A_2 = \emptyset$. We denote by B_ρ the cylinder defined by $B_\rho = \{\underline{x} : (x_1,x_2) \in A_\rho,\ 0 < x_3 < h\}$ ($\rho = 1,2$). We assume that B_ρ is occupied by an elastic material with the elasticity field $\underline{C}^{(\rho)}$ ($\rho = 1,2$), and that $\underline{C}^{(\rho)}$ is symmetric, positive definite and smooth on $\overline{B}_\rho$. Let Π_0 denote the surface of separation of the two materials. Clearly, $\Pi_0 = \{\underline{x} : (x_1,x_2) \in \Gamma_0,\ 0 \leq x_3 \leq h\}$. We can consider that the cylinder B is composed by two materials which are welded together along Π_0. Let Π_1 and Π_2 be the complementary subsets of Π defined by $\Pi_\rho = \{\underline{x} : (x_1,x_2) \in \Gamma_\rho,\ 0 \leq x_3 \leq h\}$.

Let $S^{(\rho)}(\underline{u})$ be the stress field associated with $\underline{u}$ on B_ρ. The stress-displacement relations can be written in the form

$$\underline{S}^{(\rho)}(\underline{u}) = \underline{C}^{(\rho)}[\nabla\underline{u}] \text{ on } B_\rho . \tag{4.1}$$

Assume that in the course of deformation there is no separation of material along Π_0. Thus, the displacement field and the stress vector field are continuous in passing from one medium to another. Clearly, we have the conditions

$$S^{(1)}_{i\beta}(\underline{u})\nu_\beta = S^{(2)}_{i\beta}(\underline{u})\nu_\beta \quad \text{on } \Pi_0 , \tag{4.2}$$

where $\underline{\nu}$ is the unit normal of Π_0, outward to B_1.

By an equilibrium displacement field for B we mean a vector field $\underline{u} \in C^0(B) \cap C^1(\bar{B}_1) \cap C^1(\bar{B}_2) \cap C^2(B_1) \cap C^2(B_2)$ that satisfies the conditions (4.2) and the equations of equilibrium

$$\operatorname{div} \underline{S}^{(\rho)}(\underline{u}) = \underline{0} \quad \text{on } B_\rho \quad (\rho = 1,2). \tag{4.3}$$

Let $\underline{s}^{(\rho)}(\underline{u})$ be the surface traction at regular points of ∂B_ρ, corresponding to the stress field $\underline{S}^{(\rho)}(\underline{u})$. We introduce the notations

$$\underline{R}(\underline{u}) = \sum_{\rho=1}^{2} \int_{A^0_\rho} \underline{s}^{(\rho)}(\underline{u})da , \quad \underline{H}(\underline{u}) = \sum_{\rho=1}^{2} \int_{A^0_\rho} \underline{x} \times \underline{s}^{(\rho)}(\underline{u})da, \tag{4.4}$$

where $A^0_\rho = \{\underline{x} : (x_1,x_2) \in A_\rho , \ x_3 = 0\}$.

By a solution of the relaxed Saint-Venant's problem we mean any equilibrium displacement field $\underline{u}$ that satisfies the conditions

$$\underline{s}^{(\rho)}(\underline{u}) = \underline{0} \text{ on } \Pi_\rho , \ \underline{R}(\underline{u}) = \underline{F} , \ \underline{H}(\underline{u}) = \underline{M} , \tag{4.5}$$

where $\underline{F}$ and $\underline{M}$ are prescribed vectors.

The strain energy $U(\underline{u})$ corresponding to a smooth displacement field $\underline{u}$ on $\bar{B}_1 \cup \bar{B}_2$ is

$$U(\underline{u}) = \frac{1}{2} \sum_{\rho=1}^{2} \int_{B_\rho} \nabla\underline{u}\cdot\underline{C}^{(\rho)}[\nabla\underline{u}]dv . \tag{4.6}$$

The functional $U(\cdot)$ generates the bilinear functional

$$U(\underline{u},\underline{v}) = \frac{1}{2} \sum_{\rho=1}^{2} \int_{B_\rho} \nabla\underline{u}\cdot\underline{C}^{(\rho)}[\nabla\underline{v}]dv . \tag{4.7}$$

For any smooth fields $\underline{u}$ and $\underline{v}$ on $\bar{B}_1 \cup \bar{B}_2$ we define the inner product $\langle\underline{u},\underline{v}\rangle = 2U(\underline{u},\underline{v})$. This inner product generates the energy norm

$\|\underline{u}\|_e = \langle \underline{u},\underline{u}\rangle^{1/2}$.

In view of the divergence theorem and (4.2), we find that, for any equilibrium displacement fields $\underline{u}$ and $\underline{v}$, one has

$$\langle \underline{u},\underline{v}\rangle = \sum_{\varrho=1}^{2}\int_{\Omega_\varrho} \underline{s}^{(\varrho)}(\underline{v})\cdot\underline{u}\,da\,, \tag{4.8}$$

where $\Omega_\varrho = \partial B_\varrho - \Pi_0$. It follows from (4.7) and (4.8) that

$$\sum_{\varrho=1}^{2}\int_{\Omega_\varrho} \underline{s}^{(\varrho)}(\underline{v})\cdot\underline{u}\,da = \sum_{\varrho=1}^{2}\int_{\Omega_\varrho} \underline{s}^{(\varrho)}(\underline{u})\cdot\underline{v}\,da\,. \tag{4.9}$$

We assume for the remainder of this chapter that $\underline{C}^{(\varrho)}$ is independent of the axial coordinate and belongs to C^∞.

Let V denote the set of all equilibrium displacement field $\underline{u}$ on B that satisfy the conditions $\underline{s}^{(\varrho)}(\underline{u}) = \underline{0}$ on Π_ϱ ($\varrho = 1,2$).

The next theorem can be established using the procedure given in the proof of Theorem 1.1 and (4.2).

<u>Theorem 4.1.</u> If $\underline{u}\in V$ and $\underline{u}_{,3}\in C^0(B)\cap C^1(\bar{B}_1)\cap C^1(\bar{B}_2)\cap C^2(B_1)\cap C^2(B_2)$, then $\underline{u}_{,3}\in V$ and

$$\underline{R}(\underline{u}_{,3})=\underline{0}\,,\quad H_\alpha(\underline{u}_{,3}) = \varepsilon_{\alpha\beta}R_\beta(\underline{u}),\quad H_3(\underline{u}_{,3}) = 0\,.$$

The following results are easily proven.

<u>Corollary 4.1.</u> If $\underline{u}\in K_I(F_3,M_1,M_2,M_3)$ and $\underline{u}_{,3}\in C^0(B)\cap C^1(\bar{B}_1)\cap C^1(\bar{B}_2)\cap C^2(B_1)\cap C^2(B_2)$ then $\underline{u}_{,3}\in V$ and $\underline{R}(\underline{u}_{,3}) = \underline{0}$, $\underline{H}(\underline{u}_{,3}) = \underline{0}$.

<u>Corollary 4.2.</u> If $\underline{u}\in K_{II}(F_1,F_2)$ and $\underline{u}_{,3}\in C^0(B)\cap C^1(\bar{B}_1)\cap C^1(\bar{B}_2)\cap C^2(B_1)\cap C^2(B_2)$ then $\underline{u}_{,3}\in K_I(0,F_2,-F_1,0)$.

4.2. Elastic Cylinders Composed of Nonhomogeneous and Isotropic Materials

In this section we assume that B_ϱ is occupied by an isotropic material with the Lamé moduli $\lambda^{(\varrho)}$ and $\mu^{(\varrho)}$.

α) <u>Extension, Bending and Torsion.</u> We continue to use the notation $\mathcal{R} = \{\underline{u} : \underline{u} = \underline{\alpha} + \underline{\beta}\times\underline{x}\}$ where $\underline{\alpha}$ and $\underline{\beta}$ are constant vectors. In view of

Corollary 4.1, we are led to seek a solution $\underline{v}$ of the problem (P_1) such that $\underline{v}_{,3} \in \mathcal{R}$.

Theorem 4.2. Let T be the set of all vector fields $\underline{u} \in C^0(B) \cap C^1(\bar{B}_1) \cap C^1(\bar{B}_2) \cap C^2(B_1) \cap C^2(B_2)$ such that $\underline{u}_{,3} \in \mathcal{R}$. Then there exists a vector field $\underline{v} \in T$ such that $\underline{v} \in K_I(F_3, M_1, M_2, M_3)$.

Proof. If $\underline{v} \in T$, then $\underline{v}$ has the form (1.18) where a_i (i=1,2,3,4) are arbitrary constants and $\underline{w}$ is an arbitrary vector field independent of x_3 such that $\underline{w} \in C^0(\Sigma) \cap C^1(\bar{A}_1) \cap C^1(\bar{A}_2) \cap C^2(A_1) \cap C^2(A_2)$.

The stress-displacement relations imply

$$
\begin{aligned}
S^{(\rho)}_{\alpha\beta}(\underline{v}) &= \lambda^{(\rho)}(a_\nu x_\nu + a_3)\delta_{\alpha\beta} + T^{(\rho)}_{\alpha\beta}(\underline{w}),\\
S^{(\rho)}_{3\alpha}(\underline{v}) &= \mu^{(\rho)} a_4(\varphi_{,\alpha} - \varepsilon_{\alpha\beta}x_\beta), \qquad (4.10)\\
S^{(\rho)}_{33}(\underline{v}) &= (\lambda^{(\rho)} + 2\mu^{(\rho)})(a_\alpha x_\alpha + a_3) + \lambda^{(\rho)} w_{\nu,\nu},
\end{aligned}
$$

where we have used the notations

$$
\begin{aligned}
w_3 &= a_4\varphi,\\
T^{(\rho)}_{\alpha\beta}(\underline{w}) &= \mu^{(\rho)}(w_{\alpha,\beta} + w_{\beta,\alpha}) + \lambda^{(\rho)}\delta_{\alpha\beta} w_{\nu,\nu}. \qquad (4.11)
\end{aligned}
$$

Let us prove that the functions w_i and the constants a_s (s=1,2,3,4) can be determined such that $\underline{v} \in K_I(F_3, M_1, M_2, M_3)$. The equations of equilibrium, the condition on the lateral boundary and the conditions (4.2) reduce to

$$
\begin{aligned}
(T^{(\rho)}_{\alpha\beta}(\underline{w}))_{,\beta} + f^{(\rho)}_\alpha &= 0 \text{ on } A_\rho,\\
T^{(1)}_{\alpha\beta}(\underline{w})\nu_\beta &= T^{(2)}_{\alpha\beta}(\underline{w})\nu_\beta + g_\alpha \quad \text{on } \Gamma_0, \qquad (4.12)\\
T^{(\rho)}_{\alpha\beta}(\underline{w})n_\beta &= p^{(\rho)}_\alpha \quad \text{on } \Gamma_\rho,
\end{aligned}
$$

and

$$
\begin{aligned}
(\mu^{(\rho)}\varphi_{,\alpha})_{,\alpha} &= m^{(\rho)} \text{ on } A_\rho,\\
\mu^{(1)}\frac{\partial\varphi}{\partial\nu} &= \mu^{(2)}\frac{\partial\varphi}{\partial\nu} + q \quad \text{on } \Gamma_0, \qquad (4.13)\\
\frac{\partial\varphi}{\partial\nu} &= k \quad \text{on } \Gamma.
\end{aligned}
$$

Here we have used the notations

$$f_\alpha^{(\rho)} = [\lambda^{(\rho)}(a_\eta x_\eta + a_3)]_{,\alpha}\, , \quad g_\alpha = (\lambda^{(2)}-\lambda^{(1)})(a_\eta x_\eta + a_3)\nu_\alpha \ ,$$

$$p_\alpha^{(\rho)} = -\lambda^{(\rho)}(a_\eta x_\eta + a_3)n_\alpha\, , \quad m^{(\rho)} = \varepsilon_{\alpha\beta}(x_\beta \mu^{(\rho)})_{,\alpha}\ , \tag{4.14}$$

$$q = (\mu^{(1)}-\mu^{(2)})\varepsilon_{\alpha\beta}x_\beta \nu_\alpha\, , \quad k = \varepsilon_{\alpha\beta}x_\beta n_\alpha\, .$$

The relations (4.11) and (4.12) constitute a two-dimensional boundary-value problem for the unknown functions w_α.

From the general theory developed by Fichera([37], Section 13) it follows that under suitable smoothness hypotheses on the arcs which compose Γ a solution of the boundary-value problem (4.11),(4.12) exists if and only if

$$\sum_{\rho=1}^{2}\Big[\int_{A_\rho} f_\alpha^{(\rho)}\, da + \int_{\Gamma_\rho} p_\alpha^{(\rho)} ds\Big] + \int_{\Gamma_0} g_\alpha ds = 0\ ,$$

$$\sum_{\rho=1}^{2}\Big[\int_{A_\rho} \varepsilon_{\alpha\beta}x_\alpha f_\beta^{(\rho)} da + \int_{\Gamma_\rho} \varepsilon_{\alpha\beta}x_\alpha p_\beta^{(\rho)} ds\Big] + \int_{\Gamma_0}\varepsilon_{\alpha\beta}x_\alpha g_\beta ds = 0. \tag{4.15}$$

In what follows we assume that the requirements which insure this result are fulfilled.

By the divergence theorem and (4.14) we conclude that the conditions (4.15) are satisfied. Thus, the boundary-value problem (4.12) has solutions for any constants a_1, a_2 and a_3. Clearly, we have

$$w_\alpha = \sum_{i=1}^{3} a_i\, w_\alpha^{(i)}\ ,$$

where $\underline{w}^{(i)} = (w_1^{(i)}, w_2^{(i)}, 0)$ are plane displacement fields characterized by the boundary-value problems

$$(T_{\alpha\beta}^{(\rho)}(\underline{w}^{(i)}))_{,\beta} + f_{\alpha i}^{(\rho)} = 0 \text{ on } A_\rho\ ,$$

$$T_{\alpha\beta}^{(1)}(\underline{w}^{(i)})\nu_\beta = T_{\alpha\beta}^{(2)}(\underline{w}^{(i)})\nu_\beta + g_{\alpha i} \quad \text{on } \Gamma_0, \tag{4.16}$$

$$T_{\alpha\beta}^{(\rho)}(\underline{w}^{(i)})n_\beta = p_{\alpha i}^{(\rho)} \quad \text{on } \Gamma_\rho\ ,$$

where

$$f_{\alpha\eta}^{(\rho)} = (\lambda^{(\rho)}x_\eta)_{,\alpha}\, , \quad f_{\alpha 3}^{(\rho)} = \lambda_{,\alpha}^{(\rho)}\ ,$$

$$g_{\alpha\eta} = (\lambda^{(2)} - \lambda^{(1)}) x_\eta \nu_\alpha , \quad g_{\alpha 3} = (\lambda^{(2)} - \lambda^{(1)}) \nu_\alpha ,$$

$$p^{(\rho)}_{\alpha\eta} = - \lambda^{(\rho)} x_\eta n_\alpha , \quad p^{(\rho)}_{\alpha 3} = - \lambda^{(\rho)} n_\alpha .$$

Necessary and sufficient condition to solve the boundary-value problem (4.13) is (cf. [110],[92],[37])

$$\sum_{\rho=1}^{2} \Big[-\int_{A_\rho} m^{(\rho)} da + \int_{\Gamma_\rho} k\mu^{(\rho)} ds \Big] + \int_{\Gamma_0} q \, ds = 0 . \qquad (4.17)$$

It is a simple matter to verify that the condition (4.17) is satisfied. In what follows we assume that the functions $w^{(i)}_\alpha$ (i=1,2,3) and φ are known.

The vector field $\underline{v}$ can be expressed in the form (1.27) where $\underline{v}^{(j)}$ (j=1,2,3,4) are defined by (1.28). The functions $w^{(i)}_\alpha$ and φ are characterized now by (4.16) and (4.13), respectively. Clearly, $\underline{v}^{(j)} \in V$ (j=1,2,3,4).

It follows from (4.10) that

$$\underline{S}^{(\rho)}(\underline{v}) = \sum_{j=1}^{4} a_j \, \underline{S}^{(\rho)}(\underline{v}^{(j)}) \text{ on } B_\rho ,$$

where

$$\begin{aligned}
S^{(\rho)}_{\alpha\beta}(\underline{v}^{(\eta)}) &= \lambda^{(\rho)} \delta_{\alpha\beta} x_\eta + T^{(\rho)}_{\alpha\beta}(\underline{w}^{(\eta)}), \quad (\eta = 1,2) , \\
S^{(\rho)}_{\alpha\beta}(\underline{v}^{(3)}) &= \lambda^{(\rho)} \delta_{\alpha\beta} + T^{(\rho)}_{\alpha\beta}(\underline{w}^{(3)}) , \\
S^{(\rho)}_{\alpha\beta}(\underline{v}^{(4)}) &= 0 , \quad S^{(\rho)}_{3\alpha}(\underline{v}^{(i)}) = 0 \quad (i = 1,2,3), \\
S^{(\rho)}_{3\alpha}(\underline{v}^{(4)}) &= \mu^{(\rho)}(\varphi_{,\alpha} - \varepsilon_{\alpha\beta} x_\beta), \\
S^{(\rho)}_{33}(\underline{v}^{(\eta)}) &= (\lambda^{(\rho)} + 2\mu^{(\rho)}) x_\eta + \lambda^{(\rho)} w^{(\eta)}_{\nu,\nu} \quad (\eta = 1,2), \\
S^{(\rho)}_{33}(\underline{v}^{(3)}) &= \lambda^{(\rho)} + 2\mu^{(\rho)} + \lambda^{(\rho)} w^{(3)}_{\nu,\nu} , \quad S^{(\rho)}_{33}(\underline{v}^{(4)}) = 0.
\end{aligned} \qquad (4.18)$$

It follows from (4.16) and (4.18) that

$$(S^{(\rho)}_{\alpha\beta}(\underline{v}^{(i)}))_{,\beta} = 0 \text{ on } B_\rho , \quad S^{(1)}_{\alpha\beta}(\underline{v}^{(i)}) \nu_\beta = S^{(2)}_{\alpha\beta}(\underline{v}^{(i)}) \nu_\beta \text{ on } \Pi_0,$$

$$S^{(\rho)}_{\alpha\beta}(\underline{v}^{(i)}) n_\beta = 0 \text{ on } \Pi_\rho (i=1,2,3), \quad S^{(\rho)}_{3\alpha}(\underline{v}^{(4)}))_{,\alpha} = 0 \text{ on } B_\rho , \qquad (4.19)$$

$$S^{(1)}_{3\alpha}(\underline{v}^{(4)})\nu_\alpha = S^{(2)}_{3\alpha}(\underline{v}^{(4)})\nu_\alpha \text{ on } \Pi_0, \; S^{(\rho)}_{3\alpha}(\underline{v}^{(4)})n_\alpha = 0 \text{ on } \Pi_\rho .$$

Since $\underline{v}_{,3} \in \mathcal{R}$, by Theorem 4.1 we find that

$$R_\alpha(\underline{v}) = \varepsilon_{\beta\alpha} H_\alpha(\underline{v}_{,3}) = 0 .$$

The conditions $R_3(\underline{v}) = F_3$, $\underline{H}(\underline{v}) = \underline{M}$ furnish the following system for the constants a_1, a_2, a_3 and a_4

$$J_{\alpha i}\, a_i = \varepsilon_{\alpha\beta} M_\beta , \quad J_{3i}\, a_i = -F_3 ,$$
$$D^* a_4 = -M_3 , \tag{4.20}$$

where

$$J_{\alpha i} = \sum_{\rho=1}^{2} \int_{A_\rho} x_\alpha S^{(\rho)}_{33}(\underline{v}^{(i)})da ,$$
$$J_{3i} = \sum_{\rho=1}^{2} \int_{A_\rho} S^{(\rho)}_{33}(\underline{v}^{(i)})da , \tag{4.21}$$
$$D^* = \sum_{\rho=1}^{2} \int_{A_\rho} \varepsilon_{\alpha\beta} x_\alpha S^{(\rho)}_{3\beta}(\underline{v}^{(4)})da .$$

Let us prove that the system (4.20) determines the constants a_i (i= =1,2,3,4). It follows from (4.6) and (4.8) that

$$U(\underline{v}) = \frac{1}{2} \sum_{i,j=1}^{4} < \underline{v}^{(i)}, \underline{v}^{(j)} > a_i a_j , \tag{4.22}$$

which implies

$$\det < \underline{v}^{(i)}, \underline{v}^{(j)} > \neq 0 . \tag{4.23}$$

Since $\underline{v}^{(i)} \in V$, it follows from (4.8),(4.9),(1.28) and (4.19) that

$$< \underline{v}^{(i)}, \underline{v}^{(j)} > = h\, J_{ij} , \quad < \underline{v}^{(4)}, \underline{v}^{(i)} > = 0 \; (i=1,2,3),$$

$$< \underline{v}^{(4)}, \underline{v}^{(4)} > = h\, D^* .$$

By (4.23) we conclude that

$$D^* \det (J_{ij}) \neq 0 ,$$

so that the system (4.20) can always be solved for a_1, a_2, a_3 and a_4. □

Let $\underline{v}\{\hat{a}\}$ be the solution established in the proof of Theorem 4.2. Thus we have

$$v_\alpha\{\hat{a}\} = -\frac{1}{2} a_\alpha x_3^2 - a_4 \varepsilon_{\alpha\beta} x_\beta x_3 + \sum_{i=1}^{3} a_i w_\alpha^{(i)},$$
$$v_3\{\hat{a}\} = (a_\rho x_\rho + a_3) x_3 + a_4 \varphi . \tag{4.24}$$

Note that the torsion problem can be treated independently of the extension and bending problems.

β) Flexure. Let Q be the set of all vector fields of the form (1.37), where $\hat{b}$ and $\hat{c}$ are two constant four-dimensional vectors, and $\underline{w}^*$ is a vector field independent of x_3 such that $\underline{w}^* \in C^0(\Sigma) \cap C^1(\bar{A}_1) \cap C^1(\bar{A}_2) \cap C^2(A_1) \cap C^2(A_2)$.

In view of Corollary 4.2 and Theorem 4.2 we are led to seek a solution of the flexure problem which belongs to Q.

Theorem 4.3. There exists a vector field $\underline{u}^0 \in Q$ such that $\underline{u}^0 \in K_{II}(F_1, F_2)$.

Proof. By $\underline{u}^0 \in K_{II}(F_1, F_2)$ and Corollary 4.2, $\underline{v}\{\hat{b}\} \in K_I(0, F_2, -F_1, 0)$. It follows from (4.20) that

$$J_{\alpha i}\, b_i = -F_\alpha , \quad J_{3i}\, b_i = 0 , \quad b_4 = 0 . \tag{4.25}$$

Thus, we have

$$u_\alpha^0 = -\frac{1}{6} b_\alpha x_3^3 - \frac{1}{2} c_\alpha x_3^2 - c_4 \varepsilon_{\alpha\beta} x_\beta x_3 + \sum_{j=1}^{3} (c_j + b_j x_3) w_\alpha^{(j)} + w_\alpha^*,$$

$$u_3^0 = \frac{1}{2} (b_\alpha x_\alpha + b_3) x_3^2 + (c_\alpha x_\alpha + c_3) x_3 + c_4 \varphi + \psi,$$

where we have used the notation $w_3^* = \psi$.

The stress-displacement relations imply that

$$S_{\alpha\beta}^{(\rho)}(\underline{u}^0) = T_{\alpha\beta}^{(\rho)}(\underline{w}^*) + \sum_{j=1}^{3} (c_j + b_j x_3) S_{\alpha\beta}(\underline{v}^{(j)}),$$
$$S_{\alpha 3}^{(\rho)}(\underline{u}^0) = c_4 S_{3\alpha}^{(\rho)}(\underline{v}^{(4)}) + \mu^{(\rho)}(\psi_{,\alpha} + \sum_{j=1}^{3} b_j w_\alpha^{(j)}), \tag{4.26}$$
$$S_{33}^{(\rho)}(\underline{u}^0) = \sum_{j=1}^{3} (c_j + b_j x_3) S_{33}^{(\rho)}(\underline{v}^{(j)}) + \lambda^{(\rho)} w_{\alpha,\alpha}^* .$$

Since $\underline{v}^{(i)} \in V$, the equations of equilibrium, the condition on the lateral boundary and the condition (4.2) reduce to

$$\begin{aligned}
&(T^{(\rho)}_{\alpha\beta}(\underline{w}^*))_{,\beta} = 0 \quad \text{on } A_\rho \,, \\
&T^{(1)}_{\alpha\beta}(\underline{w}^*)\nu_\beta = T^{(2)}_{\alpha\beta}(\underline{w}^*)\nu_\beta \quad \text{on } \Gamma_0 \,, \\
&T^{(\rho)}_{\alpha\beta}(\underline{w}^*)n_\beta = 0 \quad \text{on } \Gamma_\rho \,,
\end{aligned} \tag{4.27}$$

and

$$\begin{aligned}
&(\mu^{(\rho)}\psi_{,\alpha})_{,\alpha} = -\sum_{j=1}^{3} b_j[(\mu^{(\rho)}w^{(j)}_\alpha)_{,\alpha} + S^{(\rho)}_{33}(\underline{v}^{(j)})] \quad \text{on } A_\rho \,, \\
&\mu^{(1)}\frac{\partial\psi}{\partial\nu} = \mu^{(2)}\frac{\partial\psi}{\partial\nu} + (\mu^{(2)} - \mu^{(1)})\nu_\alpha \sum_{j=1}^{3} b_j w^{(j)}_\alpha \quad \text{on } \Gamma_0, \\
&\frac{\partial\psi}{\partial n} = -\sum_{j=1}^{3} b_j w^{(j)}_\alpha n_\alpha \quad \text{on } \Gamma_\rho \,.
\end{aligned} \tag{4.28}$$

It follows from (4.27) that $w^*_\alpha = 0$. A simple calculation shows that the necessary and sufficient condition for the existence of a solution to the boundary-value problem (4.28) is satisfied.

The conditions $R_\alpha(\underline{u}^0) = F_\alpha$ are satisfied on the basis of (4.25). The conditions $R_3(\underline{u}^0) = 0$, $\underline{H}(\underline{u}^0) = \underline{0}$ reduce to

$$\begin{aligned}
&J_{ij}\, c_j = 0 \,, \\
&D^* c_4 = -\sum_{\rho=1}^{2} \int_{A_\rho} \varepsilon_{\alpha\beta}\mu^{(\rho)}x_\alpha(\psi_{,\beta} + \sum_{j=1}^{3} b_j w^{(j)}_\beta)da.
\end{aligned} \tag{4.29}$$

We conclude that c_4 is determined by (4.29) and $c_i = 0$. Thus, a solution of the flexure problem which belongs to Q is given by

$$u^0_\alpha = -\frac{1}{6} b_\alpha x_3^3 - c_4\varepsilon_{\alpha\beta}x_\beta x_3 + x_3 \sum_{j=1}^{3} b_j w^{(j)}_\alpha \,,$$

$$u^0_3 = \frac{1}{2}(b_\rho x_\rho + b_3)x_3^2 + c_4\varphi + \psi. \quad \square$$

γ) Theory of Loaded Cylinders. Let $\underline{f}^{(\rho)}$ be the body force field on B_ρ. By an equilibrium displacement field on B corresponding to the body force $\underline{f}^{(\rho)}$ we mean a vector field $\underline{u} \in C^0(B) \cap C^1(\bar{B}_1) \cap C^1(\bar{B}_2) \cap C^2(B_1) \cap C^2(B_2)$ that satisfies the conditions (4.2) and the equilibrium equations

$$\operatorname{div} \underline{S}^{(\rho)}(\underline{u}) + \underline{f}^{(\rho)} = \underline{0} \quad \text{on } B_\rho \quad (\rho = 1,2) . \tag{4.30}$$

The conditions (4.5) are replaced by

$$\underline{s}^{(\rho)}(\underline{u}) = \underline{p}^{(\rho)} \text{ on } \Pi_\rho , \quad \underline{R}(\underline{u}) = \underline{F} , \quad \underline{H}(\underline{u}) = \underline{M} , \tag{4.31}$$

where $\underline{p}^{(\rho)}$, $\underline{F}$ and $\underline{M}$ are prescribed. We first consider the Almansi-Michell problem. We assume that $\underline{f}^{(\rho)}$ and $\underline{p}^{(\rho)}$ are vector fields independent of the axial coordinate which belong to C^∞.

By a solution of the Almansi-Michell problem we mean any equilibrium displacement field $\underline{u}$ on B corresponding to the body force $\underline{f}^{(\rho)}$ that satisfies the conditions (4.31).

Theorem 4.4. If $\underline{u} \in K_{III}(\underline{F},\underline{M},\underline{f}^{(\rho)},\underline{p}^{(\rho)})$, and $\underline{u}_{,3} \in C^0(B) \cap C^1(\bar{B}_1) \cap C^1(\bar{B}_2) \cap C^2(B_1) \cap C^2(B_2)$, then $\underline{u}_{,3} \in K(\underline{G},\underline{Q})$ where

$$\underline{G} = \sum_{\rho=1}^{2} \Big[\int_{\Gamma_\rho} \underline{p}^{(\rho)} ds + \int_{A_\rho} \underline{f}^{(\rho)} da \Big],$$

$$Q = \varepsilon_{\alpha\beta} \sum_{\rho=1}^{2} \Big[\int_{\Gamma_\rho} x_\beta p_3^{(\rho)} ds + \int_{A_\rho} x_\beta f_3^{(\rho)} da \Big] + \varepsilon_{\alpha\beta} F_\beta , \tag{4.32}$$

$$Q_3 = \sum_{\rho=1}^{2} \Big[\int_{\Gamma_\rho} \varepsilon_{\alpha\beta} x_\alpha p_\beta^{(\rho)} ds + \int_{A_\rho} \varepsilon_{\alpha\beta} x_\alpha f_\beta^{(\rho)} da \Big] .$$

Proof. By (4.30),

$$S_{3i}^{(\rho)}(\underline{u}_{,3}) = -(S_{\alpha i}^{(\rho)}(\underline{u}))_{,\alpha} - f_i^{(\rho)},$$

$$\varepsilon_{\alpha\beta} x_\beta S_{33}^{(\rho)}(\underline{u}_{,3}) = -\varepsilon_{\alpha\beta}(x_\beta S_{\eta 3}^{(\rho)}(\underline{u}))_{,\eta} + \varepsilon_{\alpha\eta} S_{\eta 3}^{(\rho)}(\underline{u}) - \varepsilon_{\alpha\beta} x_\beta f_3^{(\rho)} ,$$

$$\varepsilon_{\alpha\beta} x_\alpha S_{3\beta}^{(\rho)}(\underline{u}_{,3}) = -\varepsilon_{\alpha\beta}(x_\alpha S_{\eta\beta}^{(\rho)}(\underline{u}))_{,\eta} - \varepsilon_{\alpha\beta} x_\alpha f_\beta^{(\rho)}.$$

It follows from (4.4) and the divergence theorem that

$$
\begin{aligned}
R_i(\underline{u},3) &= \sum_{\rho=1}^{2}\Big[\int_{\Gamma_\rho} s_i^{(\rho)}(\underline{u})ds + \int_{A_\rho} f_i^{(\rho)}da\Big] + \\
&+ \int_{\Gamma_0}(S_{\alpha i}^{(1)}(\underline{u}) - S_{\alpha i}^{(2)}(\underline{u}))\nu_\alpha ds \quad , \\
H_\alpha(\underline{u},3) &= \varepsilon_{\alpha\beta}\sum_{\rho=1}^{2}\Big[\int_{\Gamma_\rho} x_\beta s_3^{(\rho)}(\underline{u})ds + \int_{A_\rho} x_\beta f_3^{(\rho)}da\Big] + \\
&+ \int_{\Gamma_0}\varepsilon_{\alpha\beta}x_\beta(S_{\eta 3}^{(1)}(\underline{u})-S_{\eta 3}^{(2)}(\underline{u}))\nu_\eta ds + \varepsilon_{\alpha\beta}R_\beta(\underline{u}). \\
H_3(\underline{u},3) &= \sum_{\rho=1}^{2}\Big[\int_{\Gamma_\rho}\varepsilon_{\alpha\beta}x_\alpha s_\beta^{(\rho)}(\underline{u})ds + \int_{A_\rho}\varepsilon_{\alpha\beta}x_\alpha f_\beta^{(\rho)}da\Big] + \\
&+ \int_{\Gamma_0}\varepsilon_{\alpha\beta}x_\alpha(S_{\eta\beta}^{(1)} - S_{\eta\beta}^{(2)})\nu_\eta ds \, .
\end{aligned} \tag{4.33}
$$

The desired result follows from (4.2) and (4.31). □

<u>Remark.</u> Theorem 4.4 also remains valid for anisotropic bodies where $\underline{C}^{(\rho)}$ is independent of x_3.

Let Y denote the set of all vector fields that have the form (2.4), where $\underline{v}\{\hat{a}\}$ is expressed by (4.24), $\hat{b}$, $\hat{c}$ and $\hat{d}$ are four-dimensional constant vectors, and $\underline{w}^o$ and $\underline{w}'$ are vector fields independent of x_3 such that $\underline{w}^o, \underline{w}' \in C^0(\Sigma)\cap C^1(\bar{A}_1)\cap C^1(\bar{A}_2)\cap C^2(A_1)\cap C^2(A_2)$. In view of Theorem 4.4 we are led to seek a solution of Almansi-Michell problem which belongs to Y.

<u>Theorem 4.5.</u> There exists a vector field $\underline{u}' \in Y$ such that $\underline{u}' \in K_{III}(\underline{F},\underline{M},\underline{f}^{(\rho)},\underline{p}^{(\rho)})$.

<u>Proof.</u> If $\underline{u}' \in Y$ and $\underline{u}' \in K_{III}(\underline{F},\underline{M},\underline{f}^{(\rho)},\underline{p}^{(\rho)})$, by Theorem 4.4

$$\int_0^{x_3}\underline{v}\{\hat{b}\}dx_3 + \underline{v}\{\hat{c}\} + \underline{w}^o \in K(\underline{G},\underline{Q}) \, .$$

It follows from Corollary 4.2 that $\underline{v}\{\hat{b}\} \in K_I(0,G_2,-G_1,0)$. Thus, by (4.20) we obtain

$$J_{\alpha i}\, b_i = -\sum_{\rho=1}^{2}\Big[\int_{\Gamma_\rho} p_\alpha^{(\rho)} ds + \int_{A_\rho} f_\alpha^{(\rho)} da\Big] , \tag{4.34}$$

$$J_{3i}\, b_i = 0 , \qquad b_4 = 0 .$$

Moreover, $w_\alpha^o = 0$ and $w_3^o = \psi$ where the function ψ is characterized by

$$(\mu^{(\rho)}\psi_{,\alpha})_{,\alpha} = \sum_{j=1}^{3} b_j[(\mu^{(\rho)} w_\alpha^{(j)})_{,\alpha} + S_{33}^{(\rho)}(\underline{v}^{(j)})] \text{ on } A_\rho ,$$

$$\mu^{(1)}\frac{\partial\psi}{\partial\nu} = \mu^{(2)}\frac{\partial\psi}{\partial\nu} + (\mu^{(2)} - \mu^{(1)})\nu_\alpha \sum_{j=1}^{3} b_j w_\alpha^{(j)} \text{ on } \Gamma_o ,$$

$$\frac{\partial\psi}{\partial n} = -\sum_{j=1}^{3} b_j\, w_\alpha^{(j)}\, n_\alpha \text{ on } \Gamma_\rho .$$

It follows that $\underline{u}'$ has the form (2.7) where $w_\alpha^{(i)}$ are characterized by (4.16) and φ is given by (4.13).

The stress-displacement relations imply

$$S_{\alpha\beta}^{(\rho)}(\underline{u}') = \sum_{j=1}^{3}(\tfrac{1}{2} b_j x_3^2 + c_j x_3 + d_j) S_{\alpha\beta}^{(\rho)}(\underline{v}^{(j)}) + $$
$$+ \lambda^{(\rho)}(c_4\varphi + \psi)\delta_{\alpha\beta} + T_{\alpha\beta}^{(\rho)}(\underline{w}') ,$$

$$S_{\alpha 3}^{(\rho)}(\underline{u}') = (c_4 x_3 + d_4) S_{\alpha 3}^{(\rho)}(\underline{v}^{(4)}) + $$
$$+ \mu^{(\rho)}[x_3\psi_{,\alpha} + \chi_{,\alpha} + \sum_{j=1}^{3}(b_j x_3 + c_j) w_\alpha^{(j)}] , \tag{4.35}$$

$$S_{33}^{(\rho)}(\underline{u}') = \sum_{j=1}^{3}(\tfrac{1}{2} b_j x_3^2 + c_j x_3 + d_j) S_{33}(\underline{v}^{(j)}) + $$
$$+ (\lambda^{(\rho)} + 2\mu^{(\rho)})(c_4\varphi + \psi) + \lambda^{(\rho)} w'_{\alpha,\alpha} ,$$

where $\chi = w_3'$.

It follows from $\underline{u}'_{,3} \in K(\underline{G},\underline{Q})$ that $R_3(\underline{u}') = G_3$, $\underline{H}(\underline{u}'_{,3}) = \underline{Q}$. Thus, with the help of the relations

$$S_{33}^{(\rho)}(\underline{u}',_3) = \sum_{j=1}^{3}(b_j x_3 + c_j)S_{33}^{(\rho)}(\underline{v}^{(j)}),$$

$$S_{\alpha 3}^{(\rho)}(\underline{u}',_3) = c_4\, S_{\alpha 3}^{(\rho)}(\underline{v}^{(4)}) + \mu^{(\rho)}(\psi_{,\alpha} + \sum_{i=1}^{3} b_i w_\alpha^{(i)}) , \tag{4.36}$$

we obtain the following system for the constants c_1, c_2, c_3 and c_4

$$J_{\alpha i}c_i = -\sum_{\rho=1}^{2}[\int_{\Gamma_\rho} x_\alpha p_3^{(\rho)}ds + \int_{A_\rho} x_\alpha f_3^{(\rho)}da] - F_\alpha ,$$

$$J_{3i}c_i = -\sum_{\rho=1}^{2}[\int_{\Gamma_\rho} p_3^{(\rho)}ds + \int_{A_\rho} f_3^{(\rho)}da].$$

$$D^* c_4 = -\sum_{\rho=1}^{2}[\int_{\Gamma_\rho} \varepsilon_{\alpha\beta}x_\alpha p_\beta^{(\rho)}ds + \int_{A_\rho} \varepsilon_{\alpha\beta}x_\alpha f_\beta^{(\rho)}da] - \tag{4.37}$$

$$- \sum_{\rho=1}^{2}[\int_{A_\rho} \mu^{(\rho)}\varepsilon_{\alpha\beta}x_\alpha(\psi_{,\beta} + \sum_{i=1}^{3} b_i w_\beta^{(i)})da .$$

The equations of equilibrium, the condition (4.2) and the condition on the lateral boundary reduce to

$$(T_{\alpha\beta}^{(\rho)}(\underline{w}'))_{,\beta} + h_\alpha^{(\rho)} = 0 \text{ on } A_\rho ,$$

$$T_{\alpha\beta}^{(1)}(\underline{w}')\nu_\beta = T_{\alpha\beta}^{(2)}(\underline{w}')\nu_\beta + g_\alpha \quad \text{on} \quad \Gamma_0 , \tag{4.38}$$

$$T_{\alpha\beta}^{(\rho)}(\underline{w}')n_\beta = q_\alpha^{(\rho)} \quad \text{on } \Gamma_\rho ,$$

and

$$(\mu^{(\rho)}\chi_{,\alpha})_{,\alpha} = -m^{(\rho)} \quad \text{on } A_\rho ,$$

$$\mu^{(1)}\frac{\partial\chi}{\partial\nu} = \mu^{(2)}\frac{\partial\chi}{\partial\nu} + q \text{ on } \Gamma_0 , \tag{4.39}$$

$$\mu^{(\rho)}\frac{\partial\chi}{\partial n} = k^{(\rho)} \quad \text{on } \Gamma .$$

Here we have used the following notations

$$h_\alpha^{(\rho)} = f_\alpha^{(\rho)} + [\lambda^{(\rho)}(c_4\varphi + \psi)]_{,\alpha} + \mu^{(\rho)}[(c_4\varphi_{,\alpha} + \psi_{,\alpha}) -$$
$$- c_4\varepsilon_{\alpha\beta}x_\beta + \sum_{j=1}^{3} b_j w_\alpha^{(j)}],$$

$$g_\alpha = (\lambda^{(2)} - \lambda^{(1)})(c_4\varphi + \psi)\nu_\alpha ,$$

$$q_\alpha^{(\rho)} = p_\alpha^{(\rho)} - \lambda^{(\rho)}(c_4\varphi + \psi)n_\alpha , \tag{4.40}$$

$$m^{(\rho)} = f_3^{(\rho)} + (\lambda^{(\rho)} + 2\mu^{(\rho)})(c_\alpha x_\alpha + c_3) +$$
$$+ \sum_{j=1}^{3} c_j[(\mu^{(\rho)} w_\alpha^{(j)})_{,\alpha} + \lambda^{(\rho)} w_{\alpha,\alpha}^{(j)}],$$

$$q = (\mu^{(2)} - \mu^{(1)}) \sum_{j=1}^{3} c_j w_\alpha^{(j)} \nu_\alpha , \quad k^{(\rho)} = p_3^{(\rho)} - \sum_{j=1}^{3} c_j w_\alpha^{(j)} n_\alpha \mu^{(\rho)} .$$

By the divergence theorem and (4.2) we find that

$$\sum_{\rho=1}^{2}[\int_{A_\rho} h_\alpha^{(\rho)} da + \int_{\Gamma_\rho} q_\alpha^{(\rho)} ds] + \int_{\Gamma_0} g_\alpha ds =$$
$$= \sum_{\rho=1}^{2}[\int_{A_\rho} f_\alpha^{(\rho)} da + \int_{\Gamma_\rho} p_\alpha^{(\rho)} ds] - R_\alpha(\underline{u}'_{,3}) = G_\alpha - R_\alpha(\underline{u}'_{,3}) ,$$

$$\sum_{\rho=1}^{2}[\int_{A_\rho} \varepsilon_{\alpha\beta}x_\alpha h_\beta^{(\rho)} da + \int_{\Gamma_\rho} \varepsilon_{\alpha\beta}x_\alpha q_\beta^{(\rho)} ds] + \int_{\Gamma_0} \varepsilon_{\alpha\beta}x_\alpha g_\beta ds =$$
$$= \sum_{\rho=1}^{2}[\int_{A_\rho} \varepsilon_{\alpha\beta}x_\alpha f_\beta^{(\rho)} da + \int_{\Gamma_\rho} \varepsilon_{\alpha\beta}x_\alpha p_\beta^{(\rho)} ds] - H_3(\underline{u}'_{,3}) = Q_3 - H_3(\underline{u}'_{,3}).$$

In view of Theorem 4.1, (4.34),(4.36) and (4.37) we have

$$G_\alpha - R_\alpha(\underline{u}'_{,3}) = G_\alpha - \varepsilon_{\beta\alpha}H_\beta(\underline{u}'_{,33}) = G_\alpha + J_{\alpha i}\, b_i = 0 ,$$

$$Q_3 - H_3(\underline{u}'_{,3}) = Q_3 + c_4 D^* + \sum_{\rho=1}^{2} \int_{A_\rho} \mu^{(\rho)}\varepsilon_{\alpha\beta}(\psi_{,\beta} + \sum_{i=1}^{3} b_i w_\alpha^{(i)}) da = 0.$$

We conclude that the necessary and sufficient conditions for the existence of a solution of the boundary-value problem (4.38) are satisfied.

By (4.37) and (4.40),

$$\sum_{\rho=1}^{2}\Big[\int_{A_\rho} m^{(\rho)}da + \int_{\Gamma_\rho} k^{(\rho)}ds\Big] + \int_{\Gamma_0} q ds = \sum_{\rho=1}^{2}\Big[\int_{A_\rho} f_3^{(\rho)}da + \int_{\Gamma_\rho} p_3^{(\rho)}ds\Big] + $$
$$+ J_{3i}\, c_i = 0.$$

Thus, the necessary and sufficient condition to solve the boundary-value problem (4.39) is satisfied.

It follows from (4.36) and (4.37) that

$$H_\alpha(\underline{u}'_{,3}) = \varepsilon_{\beta\alpha}J_{\beta i}c_i = \varepsilon_{\alpha\beta}\sum_{\rho=1}^{2}\Big[\int_{\Gamma_\rho} x_\beta p_3^{(\rho)}ds + \int_{A_\rho} x_\beta f_3^{(\rho)}da\Big] + \varepsilon_{\alpha\beta}F_\beta \,. \tag{4.41}$$

In view of (4.33),(4.2) and the conditions on the lateral boundary,

$$H_\alpha(\underline{u}'_{,3}) = \varepsilon_{\alpha\beta}\sum_{\rho=1}^{2}\Big[\int_{\Gamma_\rho} x_\beta p_3^{(\rho)}ds + \int_{A_\rho} x_\beta f_3^{(\rho)}da\Big] + \varepsilon_{\alpha\beta}R_\beta(\underline{u}') \,. \tag{4.42}$$

From (4.41) and (4.42) we conclude that $R_\alpha(\underline{u}') = F_\alpha$.
The remaining conditions $R_3(\underline{u}') = F_3$, $\underline{H}(\underline{u}') = \underline{M}$ reduce to

$$J_{\alpha i}d_i = \varepsilon_{\alpha\beta}M_\beta - \sum_{\rho=1}^{2}\int_{A_\rho} x_\alpha[(\lambda^{(\rho)}+2\mu^{(\rho)})(c_4\varphi+\psi)+\lambda^{(\rho)}w'_{\nu,\nu}]da \,,$$

$$J_{3i}d_i = -F_3 - \sum_{\rho=1}^{2}\int_{A_\rho}[(\lambda^{(\rho)}+2\mu^{(\rho)})(c_4\varphi+\psi)+\lambda^{(\rho)}w'_{\nu,\nu}]da \,, \tag{4.43}$$

$$D^* d_4 = -M_3 - \sum_{\rho=1}^{2}\int_{A_\rho}\varepsilon_{\alpha\beta}x_\alpha\mu^{(\rho)}(\chi_{,\beta} + \sum_{i=1}^{3} c_i\, w_\beta^{(i)})da \,.$$

The system (4.43) determines the constants d_1, d_2, d_3 and d_4. Thus we can find the vectors $\hat{b}, \hat{c}$ and $\hat{d}$, and the vector field $\underline{w}''$ such that $\underline{u}' \in K_{III}(\underline{F},\underline{M},\underline{f}^{(\rho)},\underline{p}^{(\rho)})$. □

We consider now the Almansi problem. Let $\underline{u}^*$ be an equilibrium displacement field on B which corresponds to the body force $\underline{f}^{(\rho)} = \underline{g}^{(\rho)}x_3^n$ and satisfies the conditions

$$\underline{s}^{(\rho)}(\underline{u}^*) = \underline{q}^{(\rho)}x_3^n \text{ on } \Pi_\rho, \quad \underline{R}(\underline{u}^*) = \underline{0}, \quad \underline{H}(\underline{u}^*) = \underline{0} \,. \tag{4.44}$$

Here $\underline{g}^{(\rho)}$ and $\underline{q}^{(\rho)}$ are prescribed vector fields independent of x_3 which belong to C^{∞}, and n is a positive integer or zero.

Let $\underline{u}''$ be an equilibrium displacement field on B which corresponds to the body force $\underline{f}^{(\rho)} = \underline{g}^{(\rho)} x_3^{n+1}$, and satisfies the conditions

$$\underline{s}^{(\rho)}(\underline{u}'') = \underline{q}^{(\rho)} x_3^{n+1} \text{ on } \Pi_\rho, \quad \underline{R}(\underline{u}'') = \underline{0}, \quad \underline{H}(\underline{u}'') = \underline{0}.$$

In Section 2.3 we proved that the problem of Almansi reduces to the problem of finding a vector field $\underline{u}''$ once the vector field $\underline{u}^*$ is determined.

As in Section 2.3, we are led to seek the vector field $\underline{u}''$ in the form (2.27) where $v\{\hat{a}\}$ has the form (4.24), $\underline{w}^{(i)}$ and φ are characterized by (4.12) and (4.13), respectively, and $\underline{w}^{0}$ is a vector field independent of x_3. We must determine the constants a_i (i=1,2,3,4) and the plane vector field $\underline{w}^{0}$.

The stress-displacement relations imply that

$$S_{\alpha\beta}^{(\rho)}(\underline{u}'') = (n+1)\Big[\int_0^{x_3} S_{\alpha\beta}^{(\rho)}(\underline{u}^*)dx_3 + T_{\alpha\beta}^{(\rho)}(\underline{w}^{0}) + \sum_{j=1}^{3} a_j S_{\alpha\beta}^{(\rho)}(\underline{v}^{(j)}) + \lambda^{(\rho)}\delta_{\alpha\beta} u_3^*(x_1,x_2,0)\Big],$$

$$S_{33}^{(\rho)}(\underline{u}'') = (n+1)\Big[\int_0^{x_3} S_{33}^{(\rho)}(\underline{u}^*) + \lambda^{(\rho)} w_{\alpha,\alpha}^{0} + \sum_{j=1}^{3} a_j S_{33}^{(\rho)}(\underline{v}^{(j)}) + (\lambda^{(\rho)}+2\mu^{(\rho)})u_3^*(x_1,x_2,0)\Big], \tag{4.45}$$

$$S_{\alpha 3}^{(\rho)}(\underline{u}'') = (n+1)\Big[\int_0^{x_3} S_{\alpha 3}^{(\rho)}(\underline{u}^*)dx_3 + a_4 S_{3\alpha}^{(\rho)}(\underline{v}^{(4)}) + \mu^{(\rho)} u_\alpha^*(x_1,x_2,0) + \mu^{(\rho)} w_{3,\alpha}^{0}\Big].$$

By (4.19) and (4.45), the equilibrium equations, the condition (4.2) and the conditions on the lateral boundary reduce to

$$\big(T_{\alpha\beta}^{(\rho)}(\underline{w}^{0})\big)_{,\beta} + h_\alpha^{(\rho)} = 0 \text{ on } A_\rho,$$

$$T_{\alpha\beta}^{(1)}(\underline{w}^{0})\nu_\beta = T_{\alpha\beta}^{(2)}(\underline{w}^{0})\nu_\beta + g_\alpha \quad \text{on} \quad \Gamma_0, \tag{4.46}$$

$$T^{(\rho)}_{\alpha\beta}(\underline{w}^0)n_\beta = m^{(\rho)}_\alpha \quad \text{on} \quad \Gamma_\rho ,$$

$$(\mu^{(\rho)} w^0_{3,\alpha})_{,\alpha} = -k^{(\rho)} \quad \text{on } A_\rho ,$$

$$\mu^{(1)} \frac{\partial w^0_3}{\partial \nu} = \mu^{(2)} \frac{\partial w^0_3}{\partial \nu} + \eta \quad \text{on} \quad \Gamma_0 , \tag{4.47}$$

$$\mu^{(\rho)} \frac{\partial w^0_3}{\partial \nu} = q^{(\rho)} \quad \text{on} \quad \Gamma_\rho ,$$

where

$$\begin{aligned}
h^{(\rho)}_\alpha &= (\lambda^{(\rho)} u^*_3(x_1,x_2,0))_{,\alpha} + (S^{(\rho)}_{3\alpha}(\underline{u}^*))(x_1,x_2,0),\\
g_\alpha &= (\lambda^{(2)}-\lambda^{(1)})\nu_\alpha u^*_3(x_1,x_2,0), \quad m^{(\rho)}_\alpha = -\lambda^{(\rho)} n_\alpha u^*_3(x_1,x_2,0),\\
k^{(\rho)} &= (\mu^{(\rho)} u^*_\alpha(x_1,x_2,0))_{,\alpha} + (S^{(\rho)}_{33}(\underline{u}^*))(x_1,x_2,0) ,\\
\eta &= (\mu^{(2)}-\mu^{(1)}) u^*_\alpha(x_1,x_2,0)\nu_\alpha , \quad q^{(\rho)} = -\mu^{(\rho)} n_\alpha u^*_\alpha(x_1,x_2,0).
\end{aligned} \tag{4.48}$$

It follows from (4.44) and (4.48) that

$$\sum_{\rho=1}^{2}\Big[\int_{A_\rho} h^{(\rho)}_\alpha da + \int_{\Gamma_\rho} m^{(\rho)}_\alpha ds\Big] + \int_{\Gamma_0} g_\alpha ds = -R_\alpha(\underline{u}^*) = 0,$$

$$\sum_{\rho=1}^{2}\Big[\int_{A_\rho} \varepsilon_{\alpha\beta} x_\alpha h^{(\rho)}_\beta da + \int_{\Gamma_\rho} \varepsilon_{\alpha\beta} x_\alpha m^{(\rho)}_\beta ds\Big] + \int_{\Gamma_0} \varepsilon_{\alpha\beta} x_\alpha g_\beta ds = -H_3(\underline{u}^*) = 0,$$

$$\sum_{\rho=1}^{2}\Big[\int_{A_\rho} k^{(\rho)} da + \int_{\Gamma_\rho} q^{(\rho)} ds\Big] + \int_{\Gamma_0} \eta\, ds = -R_3(\underline{u}^*) = 0.$$

Thus, the necessary and sufficient conditions for the existence of solutions to the boundary-value problems (4.46) and (4.47) are satisfied. In what follows we assume that w^0_i are known.

By Theorem 2.1, $R_\alpha(\underline{u}'') = \varepsilon_{\beta\alpha} H_\beta((n+1)\underline{u}^*) = 0$. The conditions $R_3(\underline{u}'') = 0$, $\underline{H}(\underline{u}'') = \underline{0}$ imply that

$$J_{\alpha i} a_i = -\sum_{\rho=1}^{2} \int_{A_\rho} x_\alpha [\lambda^{(\rho)} w^0_{\gamma,\gamma} + (\lambda^{(\rho)} + 2\mu^{(\rho)}) u^*_3(x_1,x_2,0)]\, da ,$$

$$J_{3i} a_i = -\sum_{\rho=1}^{2} \int_{A_\rho} [\lambda^{(\rho)} w^0_{\alpha,\alpha} + (\lambda^{(\rho)} + 2\mu^{(\rho)}) u^*_3(x_1,x_2,0)]\, da , \tag{4.49}$$

$$D^* a_4 = -\sum_{\rho=1}^{2} \int_{A_\rho} \mu^{(\rho)} \varepsilon_{\alpha\beta} x_\alpha [u^*_\beta(x_1, x_2, 0) + w^0_{3,\beta}] da .$$

Thus, the constants a_s (s=1,2,3,4) are defined by (4.49) and the vector field $\underline{w}^0$ is characterized by (4.46) and (4.47).

4.3. Elastic Cylinders Composed of Nonhomogeneous and Anisotropic Materials

In this section we study the relaxed Saint-Venant's problem for a cylinder composed of two different nonhomogeneous and anisotropic materials.

In view of Corollary 4.1 and Theorem 4.2, we are led to seek a solution of the problem (P_1) in the form

$$\begin{aligned} u_\alpha &= -\frac{1}{2} a_\alpha x_3^2 - a_4 \varepsilon_{\alpha\beta} x_\beta x_3 + \sum_{i=1}^{4} a_i w_\alpha^{(i)} , \\ u_3 &= (a_\rho x_\rho + a_3) x_3 + \sum_{i=1}^{4} a_i w_3^{(i)} , \end{aligned} \qquad (4.50)$$

where a_i (i=1,2,3,4) are constants and $\underline{w}^{(i)}$ (i=1,2,3,4) are vector fields which are independent of the axial coordinate. Let X denote the set of all vector fields of the form (4.50).

<u>Theorem 4.6.</u> Let B_ρ be anisotropic and assume that the elasticity field is positive definite, independent of the axial coordinate, and $\underline{C}^{(\rho)} \in C^\infty(\bar{A}_\rho)$ ($\rho = 1,2$). Then there exists a vector field $\underline{v}^0 \in X$ such that $\underline{v}^0 \in K_I(F_3, M_1, M_2, M_3)$.

<u>Proof.</u> Let $\underline{v}^0 \in X$. Let us prove that the vector field $\underline{w}^{(j)}$ (j=1,2,3,4) and the constants a_i (i=1,2,3, 4) can be determined such that $\underline{v}^0 \in K_I(F_3, M_1, M_2, M_3)$. Clearly, we must have $\underline{w}^{(i)} \in C^0(\Sigma) \cap C^1(\bar{A}_1) \cap C^1(\bar{A}_2) \cap C^2(A_1) \cap C^2(A_2)$. By (4.1) we find that

$$\begin{aligned} S_{ij}^{(\rho)}(\underline{v}^0) = {} & C_{ij33}^{(\rho)}(a_\alpha x_\alpha + a_3) - C_{ij\alpha 3}^{(\rho)} a_4 \varepsilon_{\alpha\beta} x_\beta + \\ & + \sum_{s=1}^{4} a_s T_{ij}^{(\rho)}(\underline{w}^{(s)}) \text{ on } B_\rho , \end{aligned} \qquad (4.51)$$

where

$$T_{ij}^{(\rho)}(\underline{w}^{(s)}) = C_{ijk\alpha}^{(\rho)} w_{k,\alpha}^{(s)} \quad \text{on } A_\rho \ . \tag{4.52}$$

The equations of equilibrium, the condition (4.2) and the condition on the lateral boundary are satisfied for any a_s (s=1,2,3,4) if and only if

$$\begin{aligned} &(T_{i\alpha}^{(\rho)}(w^{(s)}))_{,\alpha} + g_{is}^{(\rho)} = 0 \quad \text{on } A_\rho , \\ &T_{i\alpha}^{(1)}(\underline{w}^{(s)})\nu_\alpha = T_{i\alpha}^{(2)}(\underline{w}^{(s)})\nu_\alpha + h_{is} \quad \text{on } \Gamma_0 , \\ &T_{i\alpha}^{(\rho)}(\underline{w}^{(s)})n_\alpha = q_{is}^{(\rho)} \quad \text{on } \Gamma_\rho \ (s=1,2,3,4), \end{aligned} \tag{4.53}$$

where

$$\begin{aligned} &g_{i\beta}^{(\rho)}=(C_{i\alpha 33}^{(\rho)}x_\beta)_{,\alpha} , \quad g_{i3}^{(\rho)}=C_{i\alpha 33,\alpha}^{(\rho)} , \quad g_{i4}^{(\rho)}= \varepsilon_{\beta\eta}(C_{i\alpha\eta 3}^{(\rho)}x_\beta)_{,\alpha} , \\ &h_{i\beta} = (C_{i\alpha 33}^{(2)} - C_{i\alpha 33}^{(1)})x_\beta\nu_\alpha , \quad h_{i3} = (C_{i\alpha 33}^{(2)} - C_{i\alpha 33}^{(1)})\nu_\alpha , \\ &h_{i4} = \varepsilon_{\beta\eta}(C_{i\alpha\eta 3}^{(2)} - C_{i\alpha\eta 3}^{(1)})x_\beta\nu_\alpha , \quad q_{i\beta}^{(\rho)}= -C_{i\alpha 33}^{(\rho)} x_\beta n_\alpha , \\ &q_{i3}^{(\rho)} = - C_{i\alpha 33}^{(\rho)} n_\alpha , \quad q_{i4}^{(\rho)} = \varepsilon_{\eta\beta}C_{i\alpha\eta 3}^{(\rho)} x_\beta n_\alpha . \end{aligned} \tag{4.54}$$

According to Fichera [37], a solution of the boundary-value problem (4.53) exists if and only if

$$\begin{aligned} &\sum_{\rho=1}^{2}\Big[\int_{A_\rho} g_{is}^{(\rho)}da + \int_{\Gamma_\rho} q_{is}^{(\rho)}ds\Big] + \int_{\Gamma_0} h_{is}ds = 0 , \\ &\sum_{\rho=1}^{2}\Big[\int_{A_\rho}\varepsilon_{\alpha\beta}x_\alpha g_{\beta s}^{(\rho)}da + \int_{\Gamma_\rho}\varepsilon_{\alpha\beta}x_\alpha q_{\beta s}^{(\rho)}ds\Big] + \int_{\Gamma_0}\varepsilon_{\alpha\beta}x_\alpha h_{\beta s}ds = 0, \\ &\qquad (s=1,2,3,4) . \end{aligned} \tag{4.55}$$

It is a simple matter to verify that the functions $g_{is}^{(\rho)}$, $q_{is}^{(\rho)}$ and h_{is} given by (4.54) satisfy the conditions (4.55). Thus, the vector fields $\underline{w}^{(s)}$ (s=1,2,3,4) are characterized by the generalized plane strain problems (4.53).

In what follows we assume that the vector fields $\underline{w}^{(s)}$ (s=1,2,3,4) are known.

Since $\underline{v}^{o}_{,3} \in \mathcal{R}$, by Theorem 4.1 we find that $R_{\alpha}(\underline{v}^{o}) = \varepsilon_{\beta\alpha} H_{\beta}(\underline{v}^{o}_{,3}) = 0$. The conditions $R_3(\underline{v}^{o}) = F_3$, $\underline{H}(\underline{v}^{o}) = \underline{M}$ reduce to

$$\sum_{s=1}^{4} L_{\alpha s} a_s = \varepsilon_{\alpha\beta} M_{\beta}\,, \quad \sum_{s=1}^{4} L_{3s} a_s = -F_3, \quad \sum_{s=1}^{4} L_{4s} a_s = -M_3, \qquad (4.56)$$

where

$$L_{\alpha\beta} = \sum_{\rho=1}^{2} \int_{A_\rho} x_{\alpha}[C^{(\rho)}_{3333}\, x_{\beta} + T^{(\rho)}_{33}(\underline{w}^{(\beta)})]\, da\,,$$

$$L_{\alpha 3} = \sum_{\rho=1}^{2} \int_{A_\rho} x_{\alpha}[C^{(\rho)}_{3333} + T^{(\rho)}_{33}(\underline{w}^{(3)})]\, da\,,$$

$$L_{\alpha 4} = \sum_{\rho=1}^{2} \int_{A_\rho} x_{\alpha}[C^{(\rho)}_{33\eta 3}\, \varepsilon_{\beta\eta} x_{\beta} + T^{(\rho)}_{33}(\underline{w}^{(4)})]\, da\,,$$

$$L_{3\alpha} = \sum_{\rho=1}^{2} \int_{A_\rho} [C^{(\rho)}_{3333} x_{\alpha} + T^{(\rho)}_{33}(\underline{w}^{(\alpha)})]\, da\,, \qquad (4.57)$$

$$L_{33} = \sum_{\rho=1}^{2} \int_{A_\rho} [C^{(\rho)}_{3333} + T^{(\rho)}_{33}(\underline{w}^{(3)})]\, da\,,$$

$$L_{34} = \sum_{\rho=1}^{2} \int_{A_\rho} [C^{(\rho)}_{33\alpha 3}\, \varepsilon_{\beta\alpha} x_{\beta} + T^{(\rho)}_{33}(\underline{w}^{(4)})]\, da\,,$$

$$L_{4\alpha} = \sum_{\rho=1}^{2} \int_{A_\rho} \varepsilon_{\eta\beta} x_{\eta}[C^{(\rho)}_{\beta 333}\, x_{\alpha} + T^{(\rho)}_{\beta 3}(\underline{w}^{(\alpha)})]\, da\,,$$

$$L_{43} = \sum_{\rho=1}^{2} \int_{A_\rho} \varepsilon_{\eta\beta} x_{\eta}[C^{(\rho)}_{\beta 333} + T^{(\rho)}_{\beta 3}(\underline{w}^{(3)})]\, da\,,$$

$$L_{44} = \sum_{\rho=1}^{2} \int_{A_\rho} \varepsilon_{\eta\beta} x_{\eta}[C^{(\rho)}_{\beta 3\nu 3}\, \varepsilon_{\lambda\nu} x_{\lambda} + T^{(\rho)}_{\beta 3}(\underline{w}^{(4)})]\, da\,.$$

As in Section 3.3 we can prove that

$$U(\underline{v}^{o}) = \frac{1}{2}\, b \sum_{i,j=1}^{4} L_{ij}\, a_i\, a_j\,.$$

Thus,

$$\det (L_{ij}) \neq 0 \quad (i,j=1,2,3,4) \; . \tag{4.58}$$

It follows that the system (4.56) can always be solved for a_1, a_2, a_3 and a_4.

We denote by $\underline{v}^0\{\hat{a}\}$ the solution of the problem (P_1) constructed by the proof of Theorem 4.6.

Let Z denote the set of all vector fields $\underline{u}$ such that

$$\underline{u} = \int_0^{x_3} \underline{v}^0\{\hat{b}\}dx_3 + \underline{v}^0\{\hat{c}\} + \underline{w} \; , \tag{4.59}$$

where $\hat{b} = (b_1,b_2,b_3,b_4)$ and $\hat{c} = (c_1,c_2,c_3,c_4)$ are two constant four-dimensional vectors, and $\underline{w}$ is a vector field independent of x_3. In view of Corollary 4.2 and Theorem 4.2 we look for a solution of the flexure problem which belongs to Z.

Theorem 4.7. Let B_ρ be anisotropic and assume that the elasticity field $\underline{C}^{(\rho)}$ is positive definite, independent of the axial coordinate, and $\underline{C}^{(\infty)} \in C^\infty(\bar{A}_\rho)$ $(\rho = 1,2)$. Then there exists a vector field $\underline{u}^* \in Z$ such that $\underline{u}^* \in K_{II}(F_1,F_2)$.

Proof. Let $\underline{u}^* \in Z$. Let us prove that the vector field $\underline{w}$ and the constant vectors $\hat{b}$ and $\hat{c}$ can be determined such that $\underline{u}^* \in K_{II}(F_1,F_2)$. By Corollary 4.2 we obtain $\underline{v}^0\{\hat{b}\} \in K_I(0,F_2,-F_1,0)$. Then, it follows from (4.56) that

$$\sum_{i=1}^4 L_{\alpha i}b_i = -F_\alpha \; , \quad \sum_{i=1}^4 L_{3i}b_i = 0, \quad \sum_{i=1}^4 L_{4i}b_i = 0 \; . \tag{4.60}$$

This system determines the constants b_1, b_2, b_3 and b_4.

Clearly, we must have $\underline{w} \in C^0(\Sigma) \cap C^1(\bar{A}_1) \cap C^1(\bar{A}_2) \cap C^2(A_1) \cap C^2(A_2)$. It follows from (4.50) and (4.59) that

$$\begin{aligned} u_\alpha^* &= -\frac{1}{6} b_\alpha x_3^3 - \frac{1}{2} b_4 \varepsilon_{\alpha\beta} x_\beta x_3^2 - \frac{1}{2} c_\alpha x_3^2 - \\ &\quad - c_4 \varepsilon_{\alpha\beta} x_\beta x_3 + \sum_{s=1}^4 (c_s + b_s x_3) w_\alpha^{(s)} + w_\alpha \; , \\ u_3^* &= \frac{1}{2}(b_\alpha x_\alpha + b_3)x_3^2 + (c_\alpha x_\alpha + c_3)x_3 + \\ &\quad + \sum_{s=1}^4 (c_s + b_s x_3) w_\alpha^{(s)} + w_3 \; . \end{aligned} \tag{4.61}$$

Then, the stress-displacement relations imply that

$$S_{ij}^{(\rho)} = C_{ij33}^{(\rho)}[c_\alpha x_\alpha + c_3 + (b_\alpha x_\alpha + b_3)x_3] - C_{ij\alpha 3}^{(\rho)}(c_4 + b_4 x_3)\varepsilon_{\alpha\beta}x_\beta + \tag{4.62}$$

$$+ \sum_{s=1}^{4}(c_s + b_s x_3)T_{ij}^{(\rho)}(\underline{w}^{(s)}) + T_{ij}^{(\rho)}(\underline{w}) + H_{ij}^{(\rho)} \text{ on } B_\rho ,$$

where

$$T_{ij}^{(\rho)}(\underline{w}) = C_{ijk\alpha}^{(\rho)} w_{k,\alpha} ,$$

$$H_{ij}^{(\rho)} = \sum_{s=1}^{4} C_{ijk3}^{(\rho)} b_s w_k^{(s)} . \tag{4.63}$$

The equations of equilibrium, the condition (4.2) and the condition on the lateral boundary reduce to

$$(T_{i\alpha}^{(\rho)}(\underline{w}))_{,\alpha} + b_i^{(\rho)} = 0 \text{ on } B_\rho ,$$

$$T_{i\alpha}^{(1)}(\underline{w})\nu_\alpha = T_{i\alpha}^{(1)}(\underline{w})\nu_\alpha + k_i \quad \text{on} \quad \Gamma_0 , \tag{4.64}$$

$$T_{i\alpha}^{(\rho)}(\underline{w})n_\alpha = g_i^{(\rho)} \text{ on } \Gamma_\rho ,$$

where

$$b_i^{(\rho)} = C_{i333}^{(\rho)}(b_\alpha x_\alpha + b_3) - C_{i3\alpha 3}^{(\rho)} b_4 \varepsilon_{\alpha\beta} x_\beta +$$

$$+ \sum_{s=1}^{4} b_s T_{ij}^{(\rho)}(\underline{w}^{(s)}) + H_{i\alpha,\alpha}^{(\rho)} , \tag{4.65}$$

$$k_i = (H_{i\alpha}^{(2)} - H_{i\alpha}^{(1)})\nu_\alpha , \quad g_i^{(\rho)} = - H_{i\alpha}^{(\rho)} n_\alpha .$$

A simple calculation shows that the necessary and sufficient conditions for the existence of a solution of the boundary-value problem (4.64) are satisfied. Thus, the vector field $\underline{w}$ is characterized by the generalized plane strain problem (4.64).

The conditions $R_\alpha(\underline{u}^*) = F_\alpha$ are satisfied on the basis of (4.60). It follows from (4.62) that the conditions $R_3(\underline{u}^*) = 0$. $\underline{H}(\underline{u}^*) = \underline{0}$ reduce to

$$\sum_{i=1}^{4} L_{\alpha i}c_i = -\sum_{\rho=1}^{2}\int_{A_\rho} x_\alpha(H_{33}^{(\rho)} + T_{33}^{(\rho)}(\underline{w}))da ,$$

$$\sum_{i=1}^{4} L_{3i}c_i = -\sum_{\rho=1}^{2}\int_{A_\rho} (H_{33}^{(\rho)} + T_{33}^{(\rho)}(w))da , \tag{4.66}$$

$$\sum_{i=1}^{4} L_{4i}c_i = -\sum_{\rho=1}^{2}\int_{A_\rho} \varepsilon_{\alpha\beta}x_\alpha(H_{\beta 3}^{(\rho)} + T_{\beta 3}^{(\rho)})da .$$

The system (4.66) can always be solved for the constants c_i (i=1,2,3,4). □

It is easy to extend the solution to the case when B is composed of n elastic bodies with different elasticities.

4.4. Applications to the Linear Thermoelastostatics.

In what follows we use the results of Section 4.2 to study the deformation of a composed cylinder due to a prescribed temperature field. We assume that B_ρ is occupied by a homogeneous and isotropic thermoelastic material with the constitutive coefficients $\lambda^{(\rho)}, \mu^{(\rho)}$ and $\beta^{(\rho)}$. We assume that the temperature field T is independent of the axial coordinate, and that $T \in C^0(\Sigma) \cap C^\infty(\bar{A}_1) \cap C^\infty(\bar{A}_2)$. According to the body force analogy (cf. Carlson [16], Sect. 11) the thermoelastic problem reduces to the problem of finding a vector field $\underline{u} \in C^0(B) \cap C^1(\bar{B}_1) \cap C^1(\bar{B}_2) \cap C^2(B_1) \cap C^2(B_2)$ that satisfies the conditions

$$s_{\alpha\eta}^{(1)}(\underline{u})\nu_\eta = s_{\alpha\eta}^{(2)}(\underline{u})\nu_\eta - (\beta^{(2)}-\beta^{(1)})T\,\nu_\alpha ,$$
$$s_{3\eta}^{(1)}(\underline{u})\nu_\eta = s_{3\eta}^{(2)}(\underline{u})\nu_\eta \quad \text{on } \Pi_0 , \tag{4.67}$$

the equations (4.30) and the conditions (4.31), where

$$f_\alpha^{(\rho)} = -\beta^{(\rho)}T_{,\alpha} , \quad f_3^{(\rho)} = 0 ,$$

$$p_\alpha^{(\rho)} = \beta^{(\rho)}Tn_\alpha , \; p_3 = 0, \; F_\alpha = 0, \; M_3 = 0 , \tag{4.68}$$

$$F_3 = -\sum_{\rho=1}^{2}\int_{A_\rho} \beta^{(\rho)}T da, \quad M_\alpha = -\sum_{\rho=1}^{2}\int_{A_\rho} \beta^{(\rho)}\varepsilon_{\alpha\nu}x_\nu T da.$$

Let us consider a vector field $\underline{u}'' \in C^0(B) \cap C^1(\bar{B}_1) \cap C^1(\bar{B}_2) \cap C^2(B_1) \cap C^2(B_2)$ that satisfies the conditions (4.67).

It is a simple matter to see that the vector field $\underline{\omega}$ defined on B by $\omega_\alpha = u''_\alpha$, $\omega_3 = 0$ also satisfies the conditions (4.67). If $\underline{u}$ is a solution of the thermoelastic problem, then the vector field $\hat{\underline{u}}$ defined by

$$\hat{\underline{u}} = \underline{u} - \underline{\omega} ,$$

belongs to $K_{III}(\hat{\underline{F}}, \hat{\underline{M}}, \hat{\underline{f}}^{(\rho)}, \hat{\underline{p}}^{(\rho)})$, where

$$\hat{F}_\alpha = 0,\ \hat{F}_3 = F_3 + \sum_{\rho=1}^{2} \int_{A_\rho} \lambda^{(\rho)} \omega_{\alpha,\alpha} da,\ \hat{M}_3 = 0 ,$$

$$\hat{M}_\alpha = M_\alpha + \sum_{\rho=1}^{2} \int_{A_\rho} \varepsilon_{\alpha\nu} x_\nu \lambda^{(\rho)} \omega_{\gamma,\gamma} da ,\ \hat{f}^{(\rho)}_\alpha = f^{(\rho)}_\alpha + (T^{(\rho)}_{\alpha\beta}(\underline{\omega}))_{,\beta} , \tag{4.69}$$

$$\hat{f}^{(\rho)}_3 = 0,\ \hat{p}^{(\rho)}_\alpha = p^{(\rho)}_\alpha - T^{(\rho)}_{\alpha\beta}(\underline{\omega}) n_\beta ,\ \hat{p}^{(\rho)}_3 = 0 .$$

We can construct $\hat{\underline{u}}$ according to Theorem 4.4. By (4.68) and (4.69),

$$\begin{aligned} &\sum_{\rho=1}^{2} \Big[\int_{\Gamma_\rho} \hat{p}^{(\rho)}_\alpha ds + \int_{A_\rho} \hat{f}^{(\rho)}_\alpha da \Big] = 0, \\ &\sum_{\rho=1}^{2} \Big[\int_{\Gamma_\rho} \varepsilon_{\alpha\beta} x_\alpha \hat{p}^{(\rho)}_\beta ds + \int_{A_\rho} \varepsilon_{\alpha\beta} x_\alpha \hat{f}^{(\rho)}_\beta da \Big] = 0 . \end{aligned} \tag{4.70}$$

It follows from (4.34) that $b_i = 0$ (=1,2,3,4). Thus we conclude that $\psi = 0$ satisfies the boundary-value problem (4.28). From (4.37),(4.68), (4.69) and (4.70) we find that $c_i = 0$ (i=1,2,3,4). In this case $\chi = 0$ is solution of the boundary-value problem (4.39). The functions w'_α are characterized by the boundary-value problem (4.38) where

$$\begin{aligned} h^{(\rho)}_\alpha &= -\beta^{(\rho)} T_{,\alpha} + (T^{(\rho)}_{\alpha\beta}(\underline{\omega}))_{,\beta} ,\ g_\alpha = 0, \\ q^{(\rho)}_\alpha &= \beta^{(\rho)} T n_\alpha - T^{(\rho)}_{\alpha\beta}(\underline{\omega}) n_\beta . \end{aligned} \tag{4.71}$$

The system (4.43) leads to

$$J_{\alpha i} d_i = \sum_{\rho=1}^{2} \int_{A_\rho} \beta^{(\rho)} x_\alpha T da - \sum_{\rho=1}^{2} \int_{A_\rho} x_\alpha \lambda^{(\rho)} (w'_{\nu,\nu} + \omega_{\nu,\nu}) da ,$$

$$J_{3i}d_i = \sum_{\rho=1}^{2} \int_{A_\rho} \beta^{(\rho)} T da - \sum_{\rho=1}^{2} \int_{A_\rho} \lambda^{(\rho)} (w'_{\nu,\nu} + \omega_{\nu,\nu}) da , \qquad (4.72)$$

$$d_4 = 0.$$

Let $\underline{v}'$ be the vector field on B defined by

$$\underline{v}' = \underline{w}' + \underline{\omega} . \qquad (4.73)$$

It follows from (4.67) and (4.71) that $\underline{v}'$ is characterized by the following boundary-value problem

$$\begin{gathered} (T^{(\rho)}_{\alpha\beta}(\underline{v}'))_{,\beta} - \beta^{(\rho)} T_{,\alpha} = 0 \text{ on } B_\rho , \\ T^{(1)}_{\alpha\eta}(\underline{v}')\nu_\eta = T^{(2)}_{\alpha\eta}(\underline{v}')\nu_\eta - (\beta^{(2)} - \beta^{(1)}) T \nu_\alpha , \\ T^{(\rho)}_{\alpha\beta}(\underline{v}') n_\beta = \beta^{(\rho)} T n_\alpha \text{ on } \Gamma_\rho . \end{gathered} \qquad (4.74)$$

We conclude that a solution of the thermoelastic problem is given by

$$\begin{gathered} u_\alpha = -\frac{1}{2} d_\alpha x_3^2 + \sum_{s=1}^{3} d_s w_\alpha^{(i)} + v'_\alpha , \\ u_3 = (d_\alpha x_\alpha + d_3) x_3 , \end{gathered} \qquad (4.75)$$

where $w_\alpha^{(i)}$ are characterized by (4.16), v'_α are given by (4.74), and d_i (i=1,2,3) are determined by the system

$$\begin{gathered} J_{\alpha i} d_i = \sum_{\rho=1}^{2} \int_{A_\rho} x_\alpha (\beta^{(\rho)} T - \lambda^{(\rho)} v'_{\nu,\nu}) da, \\ J_{3i} d_i = \sum_{\rho=1}^{2} \int_{A_\rho} (\beta^{(\rho)} T - \lambda^{(\rho)} v'_{\nu,\nu}) da . \end{gathered} \qquad (4.76)$$

In what follows we use the above method to study the deformation of a cylinder composed by two different homogeneous and isotropic materials which is subject to the constant temperature variation T^*. We assume that the domain A_2 is bounded by two concentric circles L_1 and L_2 of radius r_1 and r_2, respectively ($r_1 < r_2$). The domain A_1 is bounded by Γ_1. We suppose that the x_3-axis passes through the center of the cross section.

Following [111], it is a simple matter to see that the general solution (4.75) applies also to this situation.

The solutions of the problems (4.16) are given by

$$w_1^{(1)} = \begin{cases} \frac{1}{2} C_1(x_1^2-x_2^2) + I_1 r^2 & \text{on } A_1, \\ \frac{1}{2}(C_2+I_2 r^{-4})(x_1^2-x_2^2)+Cr^2 & \text{on } A_2, \end{cases}$$

$$w_2^{(1)} = \begin{cases} C_1 x_1 x_2 & \text{on } A_1, \\ (C_2 + I_2 r^{-4})x_1 x_2 & \text{on } A_2, \end{cases}$$

$$w_1^{(2)} = w_1^{(2)}, \tag{4.77}$$

$$w_2^{(2)} = \begin{cases} \frac{1}{2} C_1(x_2^2-x_1^2) + I_1 r^2 & \text{on } A_1, \\ \frac{1}{2}(C_2+I_2 r^{-4})(x_2^2-x_1^2) + Cr^2 & \text{on } A_2, \end{cases}$$

$$w_\alpha^{(3)} = \begin{cases} P_1 x_\alpha & \text{on } A_1, \\ (P_2 + D_3 r^{-2})x_\alpha & \text{on } A_2, \end{cases}$$

where

$$r^2 = x_\alpha x_\alpha, \; C_\rho = G_\rho - \nu^{(\rho)}, \; P_\rho = D_\rho - \nu^{(\rho)},$$

$$G_1 = \eta^{(1)}\nu^*(r_2^4-r_1^4)d, \; G_2 = -\eta^{(2)}\nu^* r_1^4 d, \; \nu^* = \nu^{(1)} - \nu^{(2)},$$

$$I_1 = -\gamma^{(1)} \frac{G_1}{\eta^{(1)}}, \; I_2 = -\gamma^{(2)}\nu^* r_1^4 r_2^4 d, \; C = -\gamma^{(2)} \frac{G_2}{\eta^{(2)}},$$

$$\nu^{(\alpha)} = \frac{\lambda^{(\alpha)}}{2(\lambda^{(\alpha)}+\mu^{(\alpha)})}, \; \eta^{(\alpha)} = \frac{\lambda^{(\alpha)} + 3\mu^{(\alpha)}}{2\mu^{(\alpha)}(\lambda^{(\alpha)}+\mu^{(\alpha)})}, \; \gamma^{(\alpha)} = \frac{1}{2\mu^{(\alpha)}},$$

$$D_1 = \eta^* \nu^*(r_2^2-r_1^2)h_o, \; D_2 = \eta^* \nu^* r_1^2 h, \; \eta^* = \eta^{(1)} - \eta^{(2)},$$

$$D_3 = 2\gamma^{(2)}\nu^* r_1^2 r_2^2 h_o \ , \quad d^{-1} = \eta^{(2)} r_1^4 + \gamma^{(2)} r_2^4 + \eta^{(1)}(r_2^4 - r_1^4),$$

$$h_o^{-1} = q_1(r_2^2 - r_1^2) + q_2 r_1^2 + 2\gamma^{(2)} r_2^2 \ , \quad q_\rho = \eta^{(\rho)} - \gamma^{(\rho)} \ .$$

The solution of the problem (4.74) is given by

$$v'_\alpha = \begin{cases} k_1 x_\alpha & \text{on } A_1, \\ (k_2 + k_3 r^{-2}) x_\alpha & \text{on } A_2, \end{cases} \tag{4.78}$$

where

$$k_1 = \frac{1}{2} T^* h_o q_1 [q_2 \beta^{(2)}(r_2^2 - r_1^2) + q_2 \beta^{(1)} r_1^2 + 2\beta^{(1)} \gamma^{(2)} r_2^2],$$

$$k_2 = \frac{1}{2} T^* h_o q_2 [q_1 \beta^{(2)}(r_2^2 - r_1^2) + q_1 \beta^{(1)} r_1^2 + 2\beta^{(2)} \gamma^{(2)} r_2^2],$$

$$k_3 = T^* h_o \gamma^{(2)} (\beta^{(1)} q_1 - \beta^{(2)} q_2) r_1^2 r_2^2 \ .$$

It follows from (4.76) and (4.78) that

$$d_\alpha = 0, \quad d_3 = \frac{(\beta^{(1)} S_1 + \beta^{(2)} S_2) T^* - 2\lambda^{(1)} k_1 S_1 - 2\lambda^{(2)} k_2 S_2}{E^{(1)} S_1 + E^{(2)} S_2 + 2\lambda^{(1)} D_1 S_1 + 2\lambda^{(2)} D_2 S_2} \ , \tag{4.79}$$

where

$$E^{(\alpha)} = \frac{\mu^{(\alpha)}(3\lambda^{(\alpha)} + 2\mu^{(\alpha)})}{\lambda^{(\alpha)} + \mu^{(\alpha)}} \ , \quad S_1 = \pi r_1^2 \ , \quad S_2 = \pi(r_2^2 - r_1^2) \ .$$

We conclude that u_i reduce to

$$u_\alpha = d_3 w_\alpha^{(3)} + v'_\alpha \ , \qquad u_3 = d_3 x_3 \ ,$$

where $w_\alpha^{(3)}$ are given by (4.77) and d_3 is given by (4.79).

The problem of thermal stresses in a cylinder composed of two different nonhomogeneous and anisotropic materials was studied in [74].

5. SAINT-VENANT'S PROBLEM FOR COSSERAT ELASTIC BODIES

5.1. Basic Equations

The theory of Cosserat elastic bodies was extensively studied (see, for example, Toupin [145], Eringen [35], Naghdi [112], Brulin and Hsieh [15], and Nowacki [116]).

In this chapter we study the relaxed Saint-Venant's problem within the linearized theory of Cosserat elastic solids. We show that the method of Section 1.3 can be extended to derive a solution of the relaxed Saint-Venant's problem in a rational form. Minimum principles charaterizing the solutions of the problems (P_1) and (P_2) are presented. These principles suggest a solution of Truesdell's problem for Cosserat cylinders. Also included in this chapter are a study of the problems of Almansi and Michell and some illustrative applications.

Saint-Venant's problem for Cosserat elastic bodies has been studied in various papers (see, e.g. [58],[59],[153],[88],[89],[22],[131],[42]).

We assume that the body occupying B is a Cosserat elastic material. Let $\underline{u}$ denote the displacement field, and let $\underline{\varphi}$ denote the microrotation field. We denote by u the siy-dimensional vector field on B, defined by $u = (\underline{u},\underline{\varphi})=(u_1,u_2,u_3,\varphi_1,\varphi_2,\varphi_3) = (u_i,\varphi_i)$.

The strain measures associated with u are defined by (see, e.g. [116],[35])

$$e_{ij}(u) = u_{j,i} + \varepsilon_{jik}\varphi_k, \quad \varkappa_{ij}(u) = \varphi_{j,i} \ . \tag{5.1}$$

Let us note that $e_{ij}(u) = 0$, $\varkappa_{ij}(u) = 0$ if and only if $u_i = a_i + \varepsilon_{ijk}b_jx_k$, $\varphi_i = b_i$, where a_i and b_i are arbitrary constants.

Let

$$\mathcal{R}^* =\{u^o : u^o=(u_i^o, \varphi_i^o), u_i^o=a_i+\varepsilon_{ijk}b_jx_k, \varphi_i^o=b_i\}, \tag{5.2}$$

where a_i and b_i are constants. The elements of $\mathcal{R}^*$ are rigid displacements and microrotations. If $u\in\mathcal{R}^*$, then u is called a rigid motion.

The strain energy density corresponding to u is

$$W(u) = \tfrac{1}{2}A_{ijrs}e_{ij}(u)e_{rs}(u)+B_{ijrs}e_{ij}(u)\varkappa_{rs}(u)+\tfrac{1}{2}C_{ijrs}\varkappa_{ij}(u)\varkappa_{rs}(u), \quad (5.3)$$

where A_{ijrs}, B_{ijrs} and C_{ijrs} are smooth functions on $\bar{B}$ which satisfy the symmetry relations

$$A_{ijrs} = A_{rsij} \quad , \quad C_{ijrs} = C_{rsij} \quad .$$

We assume that W(u) is a positive definite quadratic form in the components of the strain measures $e_{ij}(u)$ and $\varkappa_{rs}(u)$.

The constitutive equations are given by

$$\begin{aligned} t_{ij}(u) &= A_{ijrs}e_{rs}(u) + B_{ijrs}\varkappa_{rs}(u), \\ m_{ij}(u) &= B_{rsij}e_{rs}(u) + C_{ijrs}\varkappa_{rs}(u). \end{aligned} \quad (5.4)$$

Here $t_{ij}(u)$ and $m_{ij}(u)$ are the components of the stress tensor and couple stress tensor, associated with u. For the case of an isotropic and centrosymmetric medium, the constitutive equations are

$$\begin{aligned} t_{ij}(u) &= \lambda e_{rr}(u)\delta_{ij} + (\mu+\varkappa)e_{ij}(u) + \mu e_{ji}(u), \\ m_{ij}(u) &= \alpha\varkappa_{rr}(u)\delta_{ij} + \beta\varkappa_{ji}(u) + \gamma\varkappa_{ij}(u) \ , \end{aligned} \quad (5.5)$$

where δ_{ij} is the Kronecker delta.

We call a six-dimensional vector field u an equilibrium vector field for B if $u \in C^1(\bar{B}) \cap C^2(B)$ and

$$(t_{ji}(u))_{,j} = 0, \ (m_{ji}(u))_{,j} + \varepsilon_{ijk}t_{jk}(u) = 0 \ , \quad (5.6)$$

hold on B.

Let $\underline{s}(u)$ and $\underline{q}(u)$ be the stress vector and the couple stress vector at regular points of ∂B corresponding to the stress tensor $t_{ij}(u)$ and couple stress tensor $m_{ij}(u)$ defined on $\bar{B}$, i.e.

$$s_i(u) = t_{ji}(u)n_j \ , \quad q_i(u) = m_{ji}(u)n_j \ . \quad (5.7)$$

The magnitute of the vector field $u = (u_i, \varphi_i)$ is defined by

$$|u| = (u_i u_i + \varphi_i \varphi_i)^{1/2} .$$

Let us denote $T(u) = (t_{ij}(u), m_{ij}(u))$. The magnitude of $T(u)$ is defined by

$$|T(u)| = (t_{ij}(u) t_{ij}(u) + m_{ij}(u) m_{ij}(u))^{1/2}.$$

Labelling the nine independent index combinations (ij) or $(k\ell)$ by letters $\Gamma, \Delta, \ldots$, the constitutive equations become

$$t_\Gamma(u) = A_{\Gamma\Delta} e_\Delta(u) + B_{\Gamma\Delta} \varkappa_\Delta(u), \quad m_\Gamma(u) = B_{\Delta\Gamma} e_\Delta(u) + C_{\Gamma\Delta} \varkappa_\Delta(u).$$

Now let

$$T_\Gamma = t_\Gamma, \quad T_{9+\Gamma} = m \quad , E_\Gamma = e_\Gamma, \quad E_{9+\Gamma} = \varkappa_\Gamma, \quad a_{\Gamma\Delta} = A_{\Gamma\Delta},$$

$$a_{\Gamma(9+\Delta)} = B_{\Gamma\Delta}, \quad a_{(9+\Gamma)\Delta} = B_{\Delta\Gamma}, \quad a_{(9+\Gamma)(9+\Delta)} = C_{\Gamma\Delta}.$$

Then the constitutive equations can be rewritten in the form

$$T_K(u) = a_{KL} E_L(u).$$

Here and in what follows Latin majuscules indices range over the integers $(1,2,\ldots,18)$. The strain energy density may be written as

$$W(u) = \frac{1}{2} a_{KL} E_K(u) E_L(u).$$

It follows that the tensor a_{KL} is symmetric and positive definite. Then the characteristic values a_M are all strictly positive. We call the largest characteristic value the maximum elastic moduli. Let ζ be the maximum elastic moduli. Clearly, we can write

$$2W(u) = a_M \gamma_M^2(u).$$

The characteristic values of $\underline{a}^2$ are the square of the characteristic values of $\underline{a}$. Thus, we have

$$|T(u)|^2 = T_K(u)T_K(u) = a_{KL}a_{KS}E_L(u)E_S(u) = a_M^2\gamma_M^2(u) < 2\zeta W(u). \quad (5.8)$$

The strain energy $U(u)$ corresponding to a smooth vector field u on B is

$$U(u) = \int_B W(u)dv \, . \quad (5.9)$$

In the following, two six-dimensional vector fields differing by a rigid motion will be regarded identical. The functional $U(\cdot)$ generates the bilinear functional

$$U(u,v) = \frac{1}{2}\int_B [A_{ijrs}e_{ij}(u)e_{rs}(v) + B_{ijrs}(e_{ij}(u)\varkappa_{rs}(v) + e_{ij}(v)\varkappa_{rs}(u)) + C_{ijrs}\varkappa_{ij}(u)\varkappa_{rs}(v)]dv \, . \quad (5.10)$$

The set of smooth vector fields u over $\bar{B}$ can be made into a real vector space with the inner product $\langle u,v\rangle = 2U(u,v)$. This inner product generates the energy norm $\|u\|_e^2 = \langle u,u\rangle$. For any equilibrium vector fields $u = (u_i, \varphi_i)$ and $v = (v_i, \psi_i)$ one has

$$\langle u,v\rangle = \int_{\partial B} [\underline{v}\cdot\underline{s}(u) + \underline{\psi}\cdot\underline{q}(u)]da, \quad (5.11)$$

which implies the reciprocity relation

$$\int_{\partial B} [\underline{u}\cdot\underline{s}(v) + \underline{\varphi}\cdot\underline{q}(v)]da = \int_{\partial B} [\underline{v}\cdot\underline{s}(u) + \underline{\psi}\cdot\underline{q}(u)]da \, . \quad (5.12)$$

5.2. The Relaxed Saint-Venant's Problem

Saint-Venant's problem consists in the determination of an equilibrium vector field u on B, subject to the requirements

$$\underline{s}(u) = \underline{0} \, , \quad \underline{q}(u) = \underline{0} \quad \text{on } \Pi,$$
$$\underline{s}(u) = \underline{t}^{(\alpha)} \, , \qquad \underline{q}(u) = \underline{m}^{(\alpha)} \quad \text{on } \Sigma_\alpha \, , \quad (5.13)$$

where $\underline{t}^{(\alpha)}$ and $\underline{m}^{(\alpha)}$ ($\alpha = 1,2$) are vector-valued functions preassigned on Σ . Necessary conditions for the existence of a solution to this problem are given by

$$\int_{\Sigma_1} \underline{t}^{(1)} da + \int_{\Sigma_2} \underline{t}^{(2)} da = \underline{0},$$
$$\int_{\Sigma_1} (\underline{x} \times \underline{t}^{(1)} + \underline{m}^{(1)}) da + \int_{\Sigma_2} (\underline{x} \times \underline{t}^{(2)} + \underline{m}^{(2)}) da = \underline{0}. \tag{5.14}$$

Under suitable smoothness hypotheses on the vector-valued functions $\underline{t}^{(\alpha)}$ and $\underline{m}^{(\alpha)}$, and on the curve Γ , a solution of Saint-Venant's problem exists ([50],[57],[76],[92]). This solution is determined within an arbitrary rigid motion.

In the relaxed formulation of Saint-Venant's problem, the conditions (5.13) are replaced by

$$\underline{s}(u) = \underline{0}\ , \quad \underline{q}(u) = \underline{0} \quad \text{on } \Pi,$$
$$\underline{R}(u) = \underline{F}\ , \ \underline{H}(u) = \underline{M}, \tag{5.15}$$

where $\underline{F}$ and $\underline{M}$ are prescribed vectors representing the resultant force and the resultant moment about O of the tractions acting on Σ_1. Accordingly, $\underline{R}(\cdot)$ and $\underline{H}(\cdot)$ are the vector-valued linear functionals defined by

$$R_i(u) = -\int_{\Sigma_1} t_{3i}(u) da,$$
$$H_\alpha(u) = -\int_{\Sigma_1} [\varepsilon_{\alpha\beta} x_\beta t_{33}(u) + m_{3\alpha}(u)] da\ , \tag{5.16}$$
$$H_3(u) = -\int_{\Sigma_1} [\varepsilon_{\alpha\beta} x_\alpha t_{3\beta}(u) + m_{33}(u)] da\ .$$

The necessary conditions for the existence of a solution to Saint-Venant's problem lead to the following relations

$$\int_{\Sigma_2} t_{3i}(u)\,da = -R_i(u),$$

$$\int_{\Sigma_2} [x_\alpha t_{33}(u) + \varepsilon_{\beta\alpha} m_{3\beta}(u)]\,da = \varepsilon_{\alpha\beta} H_\beta(u) - hR_\alpha(u), \tag{5.17}$$

$$\int_{\Sigma_2} [\varepsilon_{\alpha\beta} x_\alpha t_{3\beta}(u) + m_{33}(u)]\,da = -H_3(u)\,.$$

By a solution of the relaxed Saint-Venant's problem we mean any equilibrium vector field that satisfies the conditions (5.15). Let $K(\underline{F},\underline{M})$ denote the class of solutions to this problem. We continue to denote by $K_I(F_3,M_1,M_2,M_3)$ the set of all solutions of the problem (P_1) and by $K_{II}(F_1,F_2)$ the set of all solutions of the problem (P_2).

We assume for the remainder of this chapter that the material is homogeneous.

Let Λ denote the set of all equilibrium vector fields u that satisfy the conditions

$$\underline{s}(u) = \underline{0}\,,\quad \underline{q}(u) = \underline{0} \text{ on } \Pi\,.$$

<u>Theorem 5.1.</u> If $u \in \Lambda$ and $u_{,3} \in C^1(\bar{B}) \cap C^2(B)$ then $u_{,3} \in \Lambda$ and

$$\underline{R}(u_{,3}) = \underline{0}\,,\quad H_\alpha(u_{,3}) = \varepsilon_{\alpha\beta} R_\beta(u),\quad H_3(u_{,3}) = 0\,. \tag{5.18}$$

<u>Proof.</u> The first assertion follows at once from the fact that $t_{ij}(u_{,3}) = (t_{ij}(u))_{,3}$, $m_{ij}(u_{,3}) = (m_{ij}(u))_{,3}$ and hypotheses. In view of the equations (5.6) we arrive at

$$t_{3i}(u_{,3}) = (t_{3i}(u))_{,3} = -(t_{\alpha i}(u))_{,\alpha}\,,$$

$$\varepsilon_{\alpha\beta} x_\beta t_{33}(u_{,3}) + m_{3\alpha}(u_{,3}) = (m_{3\alpha}(u))_{,3} + \varepsilon_{\alpha\beta} x_\beta (t_{33}(u))_{,3} = -(m_{\rho\alpha}(u))_{,\rho} -$$

$$-\varepsilon_{\alpha ij} t_{ij}(u) - \varepsilon_{\alpha\beta} x_\beta (t_{\rho 3}(u))_{,\rho} = -(m_{\rho\alpha}(u))_{,\rho} -$$

$$-\varepsilon_{\alpha ij}t_{ij}(u)-\varepsilon_{\alpha\beta}[(x_\beta t_{\rho 3}(u)),_\rho-t_{\beta 3}]=-[m_{\rho\alpha}(u)+ +\varepsilon_{\alpha\beta}x_\beta t_{\rho 3}(u)],_\rho-\varepsilon_{\rho\alpha}t_{3\rho}(u), \qquad (5.19)$$

$$\varepsilon_{\alpha\beta}x_\alpha t_{3\beta}(u,_3)+m_{33}(u,_3)=-\varepsilon_{\alpha\beta}x_\alpha(t_{\rho\beta}(u)),_\rho-(m_{\rho 3}(u)),_\rho- -\varepsilon_{\alpha\beta}t_{\alpha\beta}(u)=-[\varepsilon_{\alpha\beta}x_\alpha t_{\rho\beta}(u)+m_{\rho 3}(u)],_\rho .$$

By the divergence theorem, (5.16) and (5.19) we conclude that

$$\underline{R}(u,_3) = \int_\Gamma \underline{s}(u)ds,$$

$$H_\alpha(u,_3) = \int_\Gamma[\varepsilon_{\alpha\beta}x_\beta s_3(u) + q_\alpha(u)]ds + \varepsilon_{\alpha\rho}R_\rho(u) ,$$

$$H_3(u,_3) = \int_\Gamma[\varepsilon_{\alpha\beta}x_\alpha s_\beta(u) + q_3(u)]ds .$$

The desired result is now immediate. □

Theorem 5.1 has the following immediate consequences:

Corollary 5.1. If $u\in K_I(F_3,M_1,M_2,M_3)$ and $u,_3\in C^1(\bar{B})\cap C^2(B)$, then $u,_3\in\Lambda$ and $\underline{R}(u,_3) = \underline{0}$, $\underline{H}(u,_3) = \underline{0}$.

Corollary 5.2. If $u\in K_{II}(F_1,F_2)$ and $u,_3\in C^1(\bar{B})\cap C^2(B)$, then $u,_3\in K_I(0,F_2,-F_1,0)$.

The above results will be used to establish a solution of the relaxed Saint-Venant's problem.

5.3. Characteristic Solutions

In the present section we establish some results necessary to extend Toupin's proof of Saint-Venant's principle to Cosserat elastic bodies.

We first study the free vibration problem in the linear theory of Cosserat elastic solids. The derivation used here follows that in [47].

If we assume null body forces and body couples, then the equations of motion are

$$(t_{ji}(u))_{,j} = \rho \ddot{u}_i \ , \ (m_{ji}(u))_{,j} + \varepsilon_{ijk} t_{jk}(u) = J_{is} \ddot{\varphi}_s \ , \tag{5.20}$$

where ρ is the density in the reference configuration, J_{rs} is the microinertia tensor, and a superposed dot denotes the material time derivative. Suppose that the density field is strictly positive and the microinertia tensor field symmetric and positive definite. For homogeneous bodies ρ and J_{rs} are constants. If B is nonhomogeneous, then we assume that J_{rs} and ρ are smooth on $\bar{B}$.

The free vibration problem is concerned with a body undergoing motions of the form

$$\begin{aligned} \hat{\underline{u}}(\underline{x},t) &= \underline{u}(\underline{x})\sin(\omega t+\theta), \quad \hat{\underline{\varphi}}(\underline{x},t) = \underline{\varphi}(\underline{x})\sin(\omega t+\theta), \\ \hat{\underline{t}}(\underline{x},t) &= \underline{t}(\underline{x})\sin(\omega t+\theta), \quad \hat{\underline{m}}(\underline{x},t) = \underline{m}(\underline{x})\sin(\omega t+\theta), \end{aligned} \tag{5.21}$$

where ω is the frequency of vibration and θ is the phase. A vector field $u = (\underline{u}, \underline{\varphi})$ is an admissible vector field provided $\underline{u}, \underline{\varphi} \in C^1(\bar{B}) \cap C^2(B)$. Let Φ be the set of all admissible vector fields. For convenience, we introduce the notations

$$\begin{aligned} [u,v] &= \int_B (u_i v_i + \varphi_i \psi_i) dv \ , \ \|u\|^2 = [u,u], \\ \langle u,v\rangle_k &= \int_B (\rho u_i v_i + J_{ij} \varphi_i \psi_j) dv, \ \|u\|_k^2 = \langle u,u\rangle_k \ , \end{aligned} \tag{5.22}$$

where $u = (u_i, \varphi_i)$ and $v = (v_i, \psi_i)$.

Let L_s (s=1,2,...,6) be the operators on Φ defined by

$$L_i(u) = (t_{ji}(u))_{,j} \ , \ L_{3+i}u = (m_{ji}(u))_{,j} + \varepsilon_{irs} t_{rs}(u). \tag{5.23}$$

It follows from (5.20) and (5.21), that

$$L_i u + \rho\omega^2 u_i = 0, \ L_{3+i} u + \omega^2 J_{ij} \varphi_j = 0. \tag{5.24}$$

By a characteristic solution for the free vibration problem we mean an ordered pair $\{\lambda, u\}$ such that λ is a scalar, $u \in \Phi$, $u = (u_i, \varphi_i)$, and

$$L_i u + \rho \lambda u_i = 0,\ L_{3+i} u + \lambda J_{ij} \varphi_j = 0,\ \|u\|_k \neq 0, \qquad \underline{s}(u) = \underline{0},\ \ \underline{q}(u) = \underline{0} \quad \text{on } \partial B. \tag{5.25}$$

We study here only the traction problem, but other boundary conditions can be also considered. We call λ a characteristic value, u an associated characteristic vector field.

Let $u, v \in \Phi$, $u = (u_i, \varphi_i)$, $v = (v_i, \psi_i)$, and assume that $Lu = (L_i u, L_{3+i} u)$ is continuous on $\bar{B}$. By (5.23) and the divergence theorem, we find that

$$[v, Lu] = \int_{\partial B} (v_i t_i(u) + \psi_i m_i(u))\,da - \langle u, v \rangle. \tag{5.26}$$

Let $\{\lambda, u\}$ be a characteristic solution. Then (5.25) and (5.26) imply that

$$[v, Lu] = -\langle u, v \rangle. \tag{5.27}$$

By (5.22) and (5.25),

$$[v, Lu] = -\lambda \langle u, v \rangle_k. \tag{5.28}$$

It follows from (5.25),(5.27) and (5.28), that

$$\lambda = \|u\|_e^2 / \|u\|_k^2. \tag{5.29}$$

Therefore we conclude that $\lambda \geq 0$. Given λ, the corresponding frequency of vibration is given by $\omega = \sqrt{\lambda}$. Clearly, $\lambda = 0$ if and only if u is rigid. Every rigid motion u^0 with $\|u^0\|_k \neq 0$ is a characteristic vector corresponding to $\lambda = 0$.

<u>Theorem 5.2.</u> Let $\{\lambda, u\}$ and $\{\lambda^*, u^*\}$ be characteristic solutions with $\lambda \neq \lambda^*$. Then

$$\langle u, u^* \rangle_k = 0. \tag{5.30}$$

We omit the proof of this theorem. For isotropic bodies, the orthogonality of characteristic vectors was established by Anderson [2]. Let Z denote the set of all vector fields $v = (v_i, \psi_i)$ that satisfy the conditions: i) $\|v\|_k \neq 0$; ii) v_i and ψ_i are continuous on $\bar{B}$ and piecewise smooth on B; iii) $v_{i,j}$ and $\psi_{i,j}$ are piecewise continuous on $\bar{B}$.

<u>Theorem 5.3.</u> Let $u_1 \in \Phi \cap Z$, and suppose that

$$\|u_1\|_e^2 / \|u_1\|_k^2 \leq \|v\|_e^2 / \|v\|_k^2 \text{ for every } v \in Z . \tag{5.31}$$

Define λ_1 by

$$\lambda_1 = \|u_1\|_e^2 / \|u_1\|_k^2 . \tag{5.32}$$

Then $\{\lambda_1, u_1\}$ is a characteristic solution, and λ_1 is the lowest characteristic value.

<u>Proof.</u> Let $v \in Z$ and let α be a real number such that $\alpha^2 \neq 1$. It follows that $\|u_1 + \alpha v\|_k \neq 0$. Clearly, the vector field $w = u_1 + \alpha v$ belongs to Z. By (5.31), $\lambda_1 = \|u_1\|_e^2 / \|u_1\|_k^2 \leq \|w\|_e^2 / \|w\|_k^2$, and hence

$$\lambda_1 \|u_1 + \alpha v\|_k^2 \leq \|u_1 + \alpha v\|_e^2 . \tag{5.33}$$

The relation (5.33) reduces to

$$\alpha^2 (\lambda_1 \|v\|_k^2 - \|v\|_e^2) + 2\alpha (\lambda_1 \langle u_1, v\rangle_k - \langle u_1, v\rangle) \leq 0.$$

This inequality must hold for any $\alpha \neq \pm 1$, so that

$$\lambda_1 \langle u_1, v\rangle_k = \langle u_1, v\rangle . \tag{5.34}$$

In view of (5.22) and (5.26), the relation (5.34) becomes

$$[Lu_1, v] + \lambda_1 \langle u_1, v\rangle_k - \int_{\partial B} (v_i t_i(u_1) + \psi_i m_i(u_1)) da = 0 . \tag{5.35}$$

Since (5.35) holds for every continuous and piecewise smooth vector field v, by the well-known fundamental lemmas (cf. [47], Sections 7, 35), we conclude that

$$L_i u_1 + \lambda_1 u_i^{(1)} = 0 , \ L_{3+i} u_1 + \lambda_1 J_{ir} \varphi_r^{(1)} = 0 \text{ on } B,$$

$$\underline{s}(u_1) = \underline{0} , \ \underline{q}(u_1) = \underline{0} \text{ on } \partial B ,$$

where $u_1 = (u_i^{(1)}, \varphi_i^{(1)})$. Since $u_1 \in Z$, it follows that $\{\lambda_1, u_1\}$ is a characteristic solution. Let $\{\lambda^*, u^*\}$ be another characteristic solution. Then, by (5.29), $\lambda^* = \|u^*\|_e^2 / \|u^*\|_k^2$. From (5.32) we conclude that $\lambda_1 \leq \lambda^*$. □

Let $u_1, u_2, \ldots, u_n$ be piecewise continuous vector fields on $\bar{B}$. Let C_r denote the set of all piecewise continuous vector fields v on $\bar{B}$ that satisfy the conditions

$$\langle v,u_r\rangle_k = 0 \ (r=1,2,\ldots,n).$$

We introduce the notations

$$Z_1 = Z \ , \ Z_r = Z\cap C_{r-1} \qquad (r\geq 2).$$

Theorem 5.4. Let $u_1,u_2,\ldots,u_n$ belong to $\Phi\cap Z$. Suppose that, for any $r\geq 1$, $u_r\in Z_r$ and

$$\|u_r\|_e^2 \ / \ \|u_r\|^2 \leq \|v\|_e^2 \ / \ \|v\|_k^2 \quad \text{for every} \quad v\in Z_r \ .$$

Define λ_r by

$$\lambda_r = \|u_r\|_e^2 \ / \ \|u_r\|_k^2 \ .$$

Then each $\{\lambda_r,u_r\}$ is a characteristic solution and

$$0\leq\lambda_1\leq\lambda_2\leq \ \ldots \qquad .$$

The proof of this theorem, which we omit, is strictly analogous to that given in the classical theory of elasticity (see [47], Section 76). We can also show that the last minimum principle characterizes all of the solutions of the free vibration problem provided $\lambda_n \longrightarrow \infty$.

As already noted, every rigid motion u^o with $\|u^o\|_k \neq 0$ is a characteristic vector corresponding to $\lambda= 0$. Let $\bar{\lambda}$ be the lowest non-zero characteristic value. Then, according to Theorem 5.4,

$$\bar{\lambda} < \|v\|_e^2 \ / \ \|v\|_k^2 \ , \tag{5.36}$$

for every admissible vector field $v = (v_i,\psi_i)$ on B that satisfies

$$\|v\|_k \neq 0 \ , \quad \langle v,u^o\rangle_k = 0 \ . \tag{5.37}$$

We assume that the form

$$\rho v_i v_i + J_{ij}\psi_i\psi_j$$

is positive definite. Then there exists a positive number $c>0$ such that for all $\underline{x} \in B$ there holds

$$\rho v_i v_i + J_{rs} \psi_r \psi_s \geq c(v_i v_i + \psi_i \psi_i).$$

This inequality implies

$$c\|v\|^2 \leq \|v\|_k^2 .$$

It follows from (5.36) that

$$\|v\|^2 \leq \frac{1}{c\bar{\lambda}} \|v\|_e^2 , \tag{5.38}$$

for every admissible vector field v on B that satisfies the conditions (5.37). For homogeneous bodies we have

$$c = \min(\rho , J_m),$$

where J_m is the smallest characteristic value of the microinertia tensor. We call c the minimum kinetic modulus.

It follows from (5.2) and (5.22) that the condition

$$\langle v, u^o \rangle_k = 0 ,$$

which appears in (5.37) is equivalent to

$$\int_B \rho v_i dv = 0, \quad \int_B (\rho \varepsilon_{ijk} x_j v_k + J_{ir} \varphi_r) dv = 0 . \tag{5.39}$$

For future use we establish the following

<u>Lemma 5.1.</u> Let $u = (u_i, \varphi_i)$ be an admissible vector field. Then there exists a unique rigid vector field $w = (w_i, \eta_i)$ such that

$$\int_B \rho \bar{u}_i dv = 0 , \quad \int_B (\rho \varepsilon_{ijk} x_j \bar{u}_k + J_{ir} \bar{\varphi}_r) dv = 0 , \tag{5.40}$$

where $\bar{u}_i = u_i + w_i$, $\bar{\varphi}_i = \varphi_i + \eta_i$.

Proof. Suppose that such a rigid vector field w exists and let

$$w_i = a_i + \varepsilon_{ijk} b_j x_k \ , \ \eta_i = b_i \ . \tag{5.41}$$

We assume that the rectangular Cartesian coordinate frame is chosen in such a way that the origin O coincides with the centroid of B.

Since

$$\int_B x_i dv = 0 \ ,$$

we obtain

$$\int_B \rho w_i dv = a_i M \ , \tag{5.42}$$

where M is the mass of B.

Further

$$\int_B (\rho \varepsilon_{ijk} x_j w_k + J_{ir} \eta_i) = K_{ij} b_j \ , \tag{5.43}$$

where

$$K_{ij} = I_{ij} + \int_B J_{ij} dv \ , \ I_{ij} = \int_B \rho (x_s x_s \delta_{ij} - x_i x_j) dv \ .$$

It is known that the tensor of inertia I_{ij} for every region B is positive definite. Since J_{ij} is also positive definite we conclude that K_{ij} is invertible. The relations (5.41),(5.42) and (5.43) imply that if such a rigid vector field w exists, it is unique and

$$a_i = -\frac{1}{M} \int_B \rho u_i dv \ , \quad K_{ij} b_j = -\int_B (\rho \varepsilon_{ijk} x_j u_k + J_{ir} \varphi_r) dv \ . \tag{5.44}$$

Conversely, it is a simple matter to verify that w defined by (5.41) and (5.44) has all the desired properties. □

5.4. Saint-Venant's Principle.

In this section we extend Theorem 1.10 to the linearized theory of Cosserat elasticity. Let u^0 be a solution of the relaxed Saint-Venant's problem and let u' be the solution of Saint-Venant's problem. It is a simple matter to see that if we define the vector field u on B by u = = u' - u^o, then u is an equilibrium vector field that satisfies the conditions

$$\underline{s}(u) = \underline{0}\,, \quad \underline{q}(u) = \underline{0} \quad \text{on } \Pi\,,$$

$$\int_{\Sigma_\alpha} \underline{s}(u)da = \underline{0}\,, \qquad \int_{\Sigma_\alpha} [\underline{x} \times \underline{s}(u) + \underline{q}(u)]\, da = \underline{0} \quad (\alpha = 1,2). \tag{5.45}$$

Thus, u is an equilibrium vector field corresponding to null body loads and to surface loads which vanish on the lateral boundary and are self-equilibrated at each end.

Let B_z denote the cylinder defined by (1.74). We denote by $U_z(u)$ the strain energy corresponding to the vector field u on B_z, i.e.

$$U_z(u) = \int_{B_z} W(u)dv\,,$$

where W(u) is defined by (5.3).

Theorem 5.5. Assume that the Cosserat elastic cylinder B is homogeneous and anisotropic and that the strain energy density is a positive definite quadratic form in the components of the strain measures. Let u be an equilibrium vector field that satisfies the conditions (5.45). Then,

$$U_z(u) \leq U_o(u) e^{-(z-\ell)/d(\ell)} \quad (z \geq \ell),$$

where

$$d(\ell) = \sqrt{\frac{\zeta}{c\,\lambda(\ell)}}\,,$$

ζ is the maximum elastic modulus, c is the minimum kinetic modulus while $\lambda(\ell)$ is the lowest non-zero characteristic value of free vibration for a slice of the cylinder, of thickness ℓ, taken normal to its generators and that has its boundary traction-free.

Proof. By (5.11) and (5.45),

$$U_z(u) = \frac{1}{2}\int_{\partial B_z}[u_i s_i(u) + \varphi_i q_i(u)]da = \frac{1}{2}\Big\{\int_{S_{h-z}}[u_i t_{3i}(u) + \qquad (5.46)$$

$$+ \varphi_i m_{3i}(u)]da - \int_{S_z}[u_i t_{3i}(u) + \varphi_i m_{3i}(u)]da\Big\},$$

where S_z denote the cross-section located at $x_3 = z$, $u = (u_i, \varphi_i)$. As in Section 1.6 we can prove that the conditions of equilibrium and (5.45) imply that

$$\int_{S_z} \underline{s}(u)da = \underline{0}, \quad \int_{S_z}[\underline{x} \times \underline{s}(u) + \underline{q}(u)]da = \underline{0}, \qquad (5.47)$$

$$\int_{S_{h-z}} \underline{s}(u)da = \underline{0}, \quad \int_{S_{h-z}}[\underline{x} \times \underline{s}(u)+\underline{q}(u)]da = \underline{0}.$$

We introduce the vector fields $u^{(\alpha)} = (u_i^{(\alpha)}, \varphi_i^{(\alpha)})$ $(\alpha = 1,2)$ defined by

$$u_i^{(\alpha)} = u_i + a_i^{(\alpha)} + \varepsilon_{ijk}b_j^{(\alpha)}x_k, \quad \varphi_i^{(\alpha)} = \varphi_i + b_i^{(\alpha)} \quad (\alpha = 1,2), \qquad (5.48)$$

where $a_i^{(\alpha)}$ and $b_i^{(\alpha)}$ are arbitrary constants. In view of (5.47) we find that

$$U_z = \frac{1}{2}\Big\{\int_{S_{h-z}}[u_i^{(1)}t_{3i}(u) + \varphi_i^{(1)}m_{3i}(u)]da - \qquad (5.49)$$

$$- \int_{S_z}[u_i^{(2)}t_{3i}(u) + \varphi_i^{(2)}m_{3i}(u)]da\,.$$

By (5.49) and the Schwarz inequality, we obtain

$$U_z(u) \leq \frac{1}{2}\Big\{\Big(\int_{S_{h-z}} |u^{(1)}|^2 da \int_{S_{h-z}} |T(u)|^2 da\Big)^{1/2} + \qquad (5.50)$$

$$+ \Big(\int_{S_z}|u^{(2)}|^2 da \int_{S_z}|T(u)|^2 da\Big)^{1/2}\Big\},$$

where $T(u) = (t_{ij}(u), m_{ij}(u))$.

If we apply the geometric-arithmetic mean inequality to (5.50), we arrive at

$$U_z(u) \leq \frac{1}{4}\Big\{\alpha\int_{S_{h-z}} |T(u)|^2 da + \frac{1}{\alpha}\int_{S_{h-z}} |u^{(1)}|^2 da + \qquad (5.51)$$

$$+ \alpha\int_{S_z} |T(u)|^2 da + \frac{1}{\alpha}\int_{S_z} |u^{(2)}|^2 da\Big\}.$$

By (5.8) and (5.51),

$$U_z(u) \leq \frac{1}{4}\Big\{2\alpha\zeta\int_{S_{h-z}} W(u)da + \frac{1}{\alpha}\int_{S_{h-z}} |u^{(1)}|da + \qquad (5.52)$$

$$+ 2\alpha\zeta\int_{S_z} W(u)da + \frac{1}{\alpha}\int_{S_z} |u^{(2)}|^2 da\Big\}.$$

Now integrate the inequality (5.52) between the limits z and $z+\ell$, $z \in (0, \frac{h}{2} - \ell)$. This yields the inequality

$$\ell E(z,\ell) \leq \frac{1}{4}\Big\{2\alpha\zeta\int_{B^*} W(u)dv + \frac{1}{\alpha}\int_{B_1} |u^{(1)}|^2 dv + \frac{1}{\alpha}\int_{B_2} |u^{(2)}|^2 dv\Big\}, \qquad (5.53)$$

where

$$\ell E(z,\ell) = \int_z^{z+\ell} U_t(u)dt, \quad B_1 = B(h-z-\ell, h-z), \quad B_2 = B(z, z+\ell),$$

$$B(s_1,s_2) = \{\underline{x} : (x_1,x_2) \in \Sigma, \ s_1 < x_3 < s_2\}, \quad (0 \leq s_1 < s_2 \leq h), \quad B^* = B_1 \cup B_2.$$

According to Lemma 5.1, we can choose the constants $a_i^{(\rho)}, b_i^{(\rho)}$ $(\rho=1,2)$ such that

$$\int_{B_1} \rho u_i^{(1)} dv = 0, \quad \int_{B_1} (\rho\varepsilon_{ijk}x_j u_k^{(1)} + J_{ij}\varphi_j^{(1)})dv = 0,$$

$$\int_{B_2} u_i^{(2)} dv = 0 \ , \quad \int_{B_2} (\rho \varepsilon_{ijk} x_j u_k^{(2)} + J_{ij} \varphi_j^{(2)}) dv = 0.$$

Thus, by (5.38),

$$\int_{B_1} | u^{(1)} |^2 dv \leq \frac{2}{c\,\lambda(\ell)} \int_{B_1} W(u) dv \ ,$$
$$\int_{B_2} | u^{(2)} |^2 dv \leq \frac{2}{c\,\lambda(\ell)} \int_{B_2} W(u) dv \ , \tag{5.54}$$

where $\lambda(\ell)$ is the lowest non-zero characteristic value of free vibration for $B(0,\ell)$. Here we have used the relations $W(u^{(\alpha)}) = W(u)$ ($\alpha = 1,2$). It follows from (5.53) and (5.54) that

$$\ell E(z,\ell) \leq k(\alpha,\ell) \int_{B^*} W(u) dv \ , \tag{5.55}$$

where

$$k(\alpha,\ell) = \frac{1}{2} (\alpha \zeta + \frac{1}{\alpha c\,\lambda(\ell)}).$$

Clearly,

$$k(\alpha,\ell) \geq \sqrt{\frac{\zeta}{c\,\lambda(\ell)}} = d(\ell) \ ,$$

for every $\alpha > 0$. By use of the same procedure as that used to prove (1.94), from (5.55) we obtain

$$E(t_2,\ell) \leq e^{-(t_2 - t_1)/d(\ell)} E(t_1,\ell) \ , \tag{5.56}$$

for $t_2 \geq t_1$. Since $E(z,\ell)$ is the mean value of $U_z(u)$ in the interval $[z, z+\ell]$, and $U_z(u)$ is a non-increasing function of z, one has

$$U_{z+\ell}(u) \leq E(z,\ell) \leq U_z(u) \ ,$$

and (5.56) implies the desired result. □

The above result was established in [9] for an isotropic cylinder subject to self-equilibrated surface loads on one of its ends, and free of surface loads on the remainder of its boundary. Note that the proof given in [9] uses Rayleigh's principle from the classical theory of elasticity.

5.5. Plane Strain Problem

We assume for the remainder of this chapter that the material is homogeneous, isotropic and centrosymmetric. With a view toward deriving a solution of the relaxed Saint-Venant's problem, we record some results concerning the plane strain problem. The state of plane strain of the cylinder B is characterized by

$$u_\alpha = u_\alpha(x_1,x_2),\quad u_3 = 0,\quad \varphi_\alpha = 0,\quad \varphi_3 = \varphi_3(x_1,x_2). \tag{5.57}$$

These restrictions, in conjunction with the constitutive equations, imply that the stress tensor and couple stress tensor are independent of the axial coordinate. It follows from (5.1) and (5.57) that the non-zero strain measures associated with $u = (u_1,u_2,0,0,0,\varphi_3)$ are given by

$$e_{\alpha\beta}(u) = u_{\beta,\alpha} + \varepsilon_{\beta\alpha}\varphi_3\,,\quad \varkappa_{\alpha 3}(u) = \varphi_{3,\alpha}\,.$$

The constitutive equations show that non-zero components of the stress tensor and couple stress tensor are $t_{\alpha\beta}$, $m_{\alpha 3}$, t_{33} and $m_{3\alpha}$. Further

$$\begin{aligned} t_{\alpha\beta}(u) &= \lambda e_{\rho\rho}(u)\delta_{\alpha\beta} + (\mu+\varkappa)e_{\alpha\beta}(u) + \mu e_{\beta\alpha}(u),\\ m_{\alpha 3}(u) &= \gamma\varkappa_{\alpha 3}(u)\,. \end{aligned} \tag{5.58}$$

Given body force $\underline{f}$ and body couple $\underline{g}$ on B, surface force $\hat{\underline{t}}$ and surface couple $\hat{\underline{m}}$ on Π, with $\underline{f}, \underline{g}, \hat{\underline{t}}$ and $\hat{\underline{m}}$ independent of x_3 and $f_3 = 0$, $g_\alpha = 0$, $\hat{t}_3 = 0$, $\hat{m}_\alpha = 0$, the plane strain problem consists in determining the vector field $u = (u_1,u_2,0,0,0,\varphi_3)$, $u \in C^1(\bar{\Sigma})\cap C^2(\Sigma)$ which satisfies the equations of equilibrium

$$(t_{\beta\alpha}(u))_{,\beta} + f_\alpha = 0,\quad (m_{\rho 3}(u))_{,\rho} + \varepsilon_{\alpha\beta}t_{\alpha\beta}(u) + g_3 = 0 \text{ on } \Sigma \tag{5.59}$$

and the boundary conditions

$$t_{\beta\alpha}(u)n_\beta = \hat{t}_\alpha\,,\quad m_{\alpha 3}(u)n_\alpha = \hat{m}_3 \quad \text{on } \Gamma\,. \tag{5.60}$$

The functions $t_{33}(u)$ and $m_{3\alpha}(u)$ can be determined after the functions u_α and φ_3 are found.

We assume for the remainder of this chapter that the domain Σ is C^∞ smooth, and that the body loads and surface loads belong to C^∞. Following [57],[113],[92],[37], we have

Theorem 5.6. The boundary-value problem (5.59),(5.60) has solutions belonging to $C^\infty(\bar{\Sigma})$ if and only if the C^∞ functions $f_\alpha, g, \hat{t}_\alpha$ and $\hat{m}_3$ satisfy the conditions

$$\int_\Sigma f_\alpha \, da + \int_\Gamma \hat{t}_\alpha \, ds = 0 ,$$
$$\int_\Sigma (\varepsilon_{\alpha\beta} x_\alpha f_\beta + g_3) da + \int_\Gamma (\varepsilon_{\alpha\beta} x_\alpha \hat{t}_\beta + \hat{m}_3) ds = 0 . \tag{5.61}$$

5.6. Extension, Bending and Torsion

Corollary 5.1 allows us to establish a method to derive a solution to the problem (P_1). Let $\mathcal{A}^*$ be the class of solutions to the relaxed Saint-Venant's problem corresponding to $\underline{F} = 0$ and $\underline{M} = \underline{0}$. In view of (5.2) it follows that $\mathcal{R}^* \subset \mathcal{A}^*$. We note that if $u \in K_I(F_3, M_1, M_2, M_3)$ and $u_{,3} \in C^1(\bar{B}) \cap C^2(B)$, then by Corollary 5.1, $u_{,3} \in \mathcal{A}^*$. It is natural to seek a solution v of the problem (P_1) such that $v_{,3}$ is a rigid motion.

Theorem 5.7. Let S be the set of all vector fields $u \in C^1(\bar{B}) \cap C^2(B)$ such that $u_{,3} \in \mathcal{R}^*$. Then there exists a vector field $v \in S$ which is solution of the problem (P_1).

Proof. Let $v \in C^1(\bar{B}) \cap C^2(B)$, $v = (v_i, \omega_i)$, such that

$$v_{,3} = (\alpha_i + \varepsilon_{ijk} \beta_j x_k, \beta_i),$$

where α_i and β_i are constants. Then it follows that

$$v_\alpha = -\frac{1}{2} a_\alpha x_3^2 - a_4 \varepsilon_{\alpha\beta} x_\beta x_3 + w_\alpha(x_1, x_2),$$

$$v_3 = (a_\rho x_\rho + a_3) x_3 + w_3(x_1, x_2), \tag{5.62}$$

$$\omega_\alpha = \varepsilon_{\alpha\beta} a_\beta x_3 + \chi_\alpha(x_1, x_2), \quad \omega_3 = a_4 x_3 + \chi_3(x_1, x_2),$$

except for an additive rigid motion. Here $w = (w_i, \chi_i)$ is an arbitrary vector field independent of x_3, and we have used the notations $a_\alpha = \varepsilon_{\rho\alpha}\beta_\rho$, $a_3 = \alpha_3$, $a_4 = \beta_3$. Let us prove that the functions w_i, χ_i and the constants a_s (s=1,2,3,4) can be determined so that $v \in K_I(F_3, M_1, M_2, M_3)$. It follows from (5.1) and (5.62) that

$$e_{\alpha\beta}(v) = e_{\alpha\beta}(w^0), \quad e_{3\alpha}(v) = -\varepsilon_{\alpha\beta}(a_4 x_\beta + \chi_\beta),$$

$$e_{\alpha 3}(v) = e_{\alpha 3}(w'), \quad e_{33}(v) = a_\rho x_\rho + a_3,$$

$$\varkappa_{\alpha\beta}(v) = \varkappa_{\alpha\beta}(w'), \quad \varkappa_{3\alpha}(v) = \varepsilon_{\alpha\beta} a_\beta,$$

$$\varkappa_{\alpha 3}(v) = \varkappa_{\alpha 3}(w^0), \quad \varkappa_{33}(v) = a_4,$$

where

$$w^0 = (w_1, w_2, 0, 0, 0, \chi_3), \quad w' = (0, 0, w_3, \chi_1, \chi_2, 0). \tag{5.63}$$

Next, a simple calculation shows that

$$t_{\alpha\beta}(v) = \lambda(a_\rho x_\rho + a_3)\delta_{\alpha\beta} + T_{\alpha\beta}(w^0), \quad t_{\alpha 3}(v) = P_\alpha(w') - \mu a_4 \varepsilon_{\alpha\rho} x_\rho,$$

$$t_{3\alpha}(v) = Q_\alpha(w') + (\mu + \varkappa) a_4 \varepsilon_{\beta\alpha} x_\beta,$$

$$t_{33}(v) = (\lambda + 2\mu + \varkappa)(a_\rho x_\rho + a_3) + \lambda e_{\rho\rho}(w^0), \tag{5.64}$$

$$m_{\nu\eta}(v) = \alpha a_4 \delta_{\nu\eta} + H_{\nu\eta}(w'), \quad m_{\alpha 3}(v) = \beta \varepsilon_{\alpha\rho} a_\rho + M_{\alpha 3}(w^0),$$

$$m_{3\alpha}(v) = \gamma \varepsilon_{\alpha\rho} a_\rho + \beta \chi_{3,\alpha}, \quad m_{33}(v) = (\alpha + \beta + \gamma) a_4 + \alpha \chi_{\rho,\rho},$$

where

$$T_{\alpha\beta}(w^0)=\lambda e_{\varrho\varrho}(w^0)\delta_{\alpha\beta}+(\mu+\varkappa)e_{\alpha\beta}(w^0)+\mu e_{\beta\alpha}(w^0),$$
$$M_{\alpha 3}(w^0)=\gamma\varkappa_{\alpha 3}(w^0),\quad P_{\alpha}(w')=(\mu+\varkappa)w_{3,\alpha}+\varkappa\varepsilon_{\alpha\beta}\chi_{\beta}, \tag{5.65}$$
$$Q_{\alpha}(w')=\mu w_{3,\alpha}+\varkappa\varepsilon_{\beta\alpha}\chi_{\beta},\quad H_{\nu\eta}(w')=\alpha\chi_{\varrho,\varrho}\delta_{\eta\nu}+\beta\chi_{\nu,\eta}+\gamma\chi_{\eta,\nu}.$$

For convenience, we introduce the following notations

$$w_3 = a_4\varphi,\quad \chi_{\alpha}=a_4\psi_{\alpha},\quad \hat{w}=(0,0,\varphi,\psi_1,\psi_2,0)\,. \tag{5.66}$$

Clearly, $w = w^0 + a_4\hat{w}$. Let $\mathcal{T}$ be the set of all vector fields $\hat{w}\in C^1(\bar{B})\cap C^2(B)$ such that $\hat{w}=(0,0,\varphi,\psi_1,\psi_2,0)$. We introduce the operators L_i on $\mathcal{T}$ defined by

$$L_{\nu}\hat{w}=\gamma\Delta\psi_{\nu}+(\alpha+\beta)\psi_{\varrho,\varrho\nu}+\varkappa\varepsilon_{\nu\beta}\varphi_{,\beta}-2\varkappa\psi_{\nu}, \tag{5.67}$$
$$L_3\hat{w}=(\mu+\varkappa)\Delta\varphi+\varkappa\varepsilon_{\alpha\beta}\psi_{\beta,\alpha}.$$

In view of (5.64),(5.66) and (5.67), the equations of equilibrium and the conditions on the lateral boundary reduce to

$$(T_{\beta\alpha}(w^o))_{,\beta}+f^o_{\alpha}=0,\quad (M_{\varrho 3}(w^0))_{,\varrho}+\varepsilon_{\alpha\beta}T_{\alpha\beta}(w^o)=0 \text{ on } \Sigma, \tag{5.68}$$
$$T_{\beta\alpha}(w^o)n_{\beta}=t^o_{\alpha},\quad M_{\alpha 3}(w^o)n_{\alpha}=m^o_3 \quad\text{on } \Gamma,$$

and

$$L_i\hat{w}=h_i \text{ on } \Sigma,\quad N_i\hat{w}=\zeta_i \quad\text{on } \Gamma, \tag{5.69}$$

where

$$f^0_{\alpha}=\lambda a_{\alpha},\quad t^0_{\alpha}=-\lambda(a_{\varrho}x_{\varrho}+a_3)n_{\alpha},\quad m^0_3=\beta\varepsilon_{\varrho\alpha}a_{\varrho}n_{\alpha}, \tag{5.70}$$
$$h_{\alpha}=\varkappa x_{\alpha},\quad h_3=0,\quad \zeta_{\nu}=-\alpha n_{\nu},\quad \zeta_3=\mu\varepsilon_{\alpha\beta}x_{\beta}n_{\alpha},$$

and

$$N_{\nu}\hat{w}=(\alpha\psi_{\varrho,\varrho}\delta_{\eta\nu}+\beta\psi_{\eta,\nu}+\gamma\psi_{\nu,\eta})n_{\eta}, \tag{5.71}$$
$$N_3\hat{w}=(\mu+\varkappa)\frac{\partial\varphi}{\partial n}+\varkappa\varepsilon_{\alpha\beta}\psi_{\beta}n_{\alpha}.$$

It follows from (5.63),(5.65),and (5.68) that w^o is characterized by a plane strain problem (cf. Sect. 5.5). It is a simple matter to verify that the necessary and sufficient conditions to solve the boundary-value problem (5.68) are satisfied. Thus, the boundary-value problem (5.68) has solutions for any constants a_1, a_2 and a_3. We denote by $w^{(i)} = (u_1^{(i)}, u_2^{(i)}, 0, 0, 0, \varphi_3^{(i)})$ (i=1,2,3) a solution of the boundary-value problem (5.68) when $a_j = \delta_{ij}$. Clearly, we have

$$w^o = \sum_{i=1}^{3} a_i \, w^{(i)} , \tag{5.72}$$

where $w^{(i)}$ are characterized by the equations

$$\begin{aligned} &(T_{\beta\alpha}(w^{(\rho)}))_{,\beta} + \lambda\delta_{\alpha\rho} = 0 , \quad (T_{\beta\alpha}(w^{(3)}))_{,\beta} = 0 , \\ &(M_{\rho 3}(w^{(i)}))_{,\rho} + \varepsilon_{\alpha\beta} T_{\alpha\beta}(w^{(i)}) = 0 \quad \text{on } \Sigma , \end{aligned} \tag{5.73}$$

and the boundary conditions

$$\begin{aligned} &T_{\beta\alpha}(w^{(\rho)}) n_\beta = -\lambda x_\rho n_\alpha , \quad T_{\beta\alpha}(w^{(3)}) n_\beta = -\lambda n_\alpha , \\ &M_{\alpha 3}(w^{(\rho)}) n_\alpha = \beta \varepsilon_{\rho\alpha} n_\alpha , \quad M_{\alpha 3}(w^{(3)}) n_\alpha = 0 \quad \text{on } \Gamma . \end{aligned} \tag{5.74}$$

In what follows we assume that the vector fields $w^{(i)}$ (i=1,2,3) are known.

Let us consider now the boundary-value problem

$$L_i \hat{w} = \eta_i \text{ on } \Sigma , \quad N_i \hat{w} = \rho_i \text{ on } \Gamma , \tag{5.75}$$

where η_i and ρ_i are C^∞ functions. In [58],[72] we established that the boundary-value problem (5.75) has a solution $\hat{w} \in C^1(\bar{\Sigma}) \cap C^2(\Sigma)$ if and only if

$$\int_\Sigma \eta_3 da - \int_\Gamma \rho_3 ds = 0 . \tag{5.76}$$

It is a simple matter to see that the necessary and sufficient for the existence of a solution to the boundary-value problem (5.69) is satisfied.

It follows from (5.62),(5.63),(5.66) and (5.72) that the vector field v can be written in the form

$$v = \sum_{j=1}^{4} a_j \, v^{(j)} \;, \tag{5.77}$$

where the vector fields $v^{(j)} = (v_i^{(j)}, \omega_i^{(j)})$ (j=1,2,3,4) are defined by

$$\begin{aligned} & v_\alpha^{(\beta)} = -\tfrac{1}{2} x_3^2 \delta_{\alpha\beta} + u_\alpha^{(\beta)}, \; v_\alpha^{(3)} = u_\alpha^{(3)}, \; v_\alpha^{(4)} = \varepsilon_{\beta\alpha} x_\beta x_3, \\ & v_3^{(\beta)} = x_\beta x_3, \; v_3^{(3)} = x_3, \; v_3^{(4)} = \varphi \,, \; \omega_\alpha^{(\beta)} = \varepsilon_{\alpha\beta} x_3, \\ & \omega_\alpha^{(3)} = 0 \,, \; \omega_\alpha^{(4)} = \psi_\alpha, \; \omega_3^{(i)} = \varphi_3^{(i)} \,, \; \omega_3^{(4)} = x_3 \,. \end{aligned} \tag{5.78}$$

We note that $v^{(j)} \in \Lambda$ (j=1,2,3,4). Clearly, (5.77) and (5.78) lead to

$$\begin{aligned} & v_\alpha = -\tfrac{1}{2} x_3^2 a_\alpha + a_4 \varepsilon_{\beta\alpha} x_\beta x_3 + \sum_{i=1}^{3} a_i u_\alpha^{(i)}, \; v_3 = (a_\rho x_\rho + a_3) x_3 + a_4 \varphi \,, \\ & \omega_\alpha = \varepsilon_{\alpha\beta} a_\beta x_3 + a_4 \psi_\alpha \,, \quad \omega_3 = a_4 x_3 + \sum_{i=1}^{3} a_i \varphi_3^{(i)}. \end{aligned} \tag{5.79}$$

By (5.64),(5.66) and (5.72) we arrive at

$$t_{ij}(v) = \sum_{s=1}^{4} a_s t_{ij}(v^{(s)}), \; m_{ij}(v) = \sum_{s=1}^{4} a_s m_{ij}(v^{(s)}), \tag{5.80}$$

where

$$\begin{aligned} & t_{\alpha\beta}(v^{(\rho)}) = \lambda x_\rho \delta_{\alpha\beta} + T_{\alpha\beta}(w^{(\rho)}), \; t_{\alpha\beta}(v^{(3)}) = \lambda \delta_{\alpha\beta} + T_{\alpha\beta}(w^{(3)}), \\ & t_{\alpha\beta}(v^{(4)}) = 0, \; t_{\alpha 3}(v^{(i)}) = 0, \; t_{\alpha 3}(v^{(4)}) = P_\alpha(\hat{w}) - \mu \varepsilon_{\alpha\beta} x_\beta, \\ & t_{3\alpha}(v^{(i)}) = 0, \; t_{3\alpha}(v^{(4)}) = Q_\alpha(\hat{w}) + (\mu + \varkappa) \varepsilon_{\beta\alpha} x_\beta \,, \\ & t_{33}(v^{(\rho)}) = (\lambda + 2\mu + \varkappa) x_\rho + \lambda u_{\alpha,\alpha}^{(\rho)}, \; t_{33}(v^{(3)}) = \lambda + 2\mu + \varkappa + \lambda u_{\alpha,\alpha}^{(3)}, \\ & t_{33}(v^{(4)}) = 0, \; m_{\nu\eta}(v^{(i)}) = 0, \; m_{\nu\eta}(v^{(4)}) = \alpha \delta_{\nu\eta} + H_{\nu\eta}(\hat{w}), \\ & m_{\alpha 3}(v^{(\rho)}) = \beta \varepsilon_{\alpha\rho} + M_{\alpha 3}(w^{(\rho)}), \; m_{\alpha 3}(v^{(3)}) = M_{\alpha 3}(w^{(3)}), \; m_{\alpha 3}(v^{(4)}) = 0, \\ & m_{3\alpha}(v^{(4)}) = 0, \; m_{33}(v^{(i)}) = 0, \; m_{33}(v^{(4)}) = \alpha + \beta + \gamma + \alpha \psi_{\rho,\rho} \,, \\ & m_{3\alpha}(v^{(\rho)}) = \gamma \varepsilon_{\alpha\rho} + \beta \varphi_{3,\alpha}^{(\rho)}, \; m_{3\alpha}(v^{(3)}) = \beta \varphi_{3,\alpha}^{(3)}, \end{aligned} \tag{5.81}$$

The conditions on the terminal cross-section Σ_1 are

$$R_\alpha(v) = 0, \ R_3(v) = F_3 \ , \ \underline{H}(v) = \underline{M} \ . \qquad (5.82)$$

Since $v_{,3} \in \mathcal{R}^*$, by Theorem 5.1 we find that $R_\alpha(v) = 0$. The remaining conditions from (5.82) furnish the following system for the constants a_1, a_2, a_3 and a_4

$$D_{\alpha j} a_j = \varepsilon_{\alpha\rho} M_\rho \ , \ D_{3j} a_j = - F_3 ,$$
$$D a_4 = - M_3 \ , \qquad (5.83)$$

where

$$D_{\alpha\beta} = \int_\Sigma \{ x_\alpha [(\lambda + 2\mu + \varkappa) x_\beta + \lambda u^{(\beta)}_{\nu,\nu}] - \beta \varepsilon_{\alpha\rho} \varphi^{(\beta)}_{3,\rho} + \gamma \delta_{\alpha\beta} \} da,$$

$$D_{\alpha 3} = \int_\Sigma \{ x_\alpha (\lambda + 2\mu + \varkappa + \lambda u^{(3)}_{\nu,\nu}) - \beta \varepsilon_{\alpha\rho} \varphi^{(3)}_{3,\rho} \} da \ , \qquad (5.84)$$

$$D_{3\alpha} = \int_\Sigma [(\lambda + 2\mu + \varkappa) x_\alpha + \lambda u^{(\alpha)}_{\nu,\nu}] da, \ D_{33} = \int_\Sigma [\lambda + 2\mu + \varkappa + \lambda u^{(3)}_{\rho,\rho}] da,$$

$$D = \int_\Sigma [\mu \varepsilon_{\alpha\beta} x_\alpha \varphi_{,\beta} + \varkappa x_\alpha \psi_\alpha + (\mu + \varkappa) x_\rho x_\rho +$$
$$+ \alpha \psi_{\rho,\rho} + \alpha + \beta + \gamma] \, da \ .$$

Clearly, the constants D_{ij} and D can be calculated after the functions $\{ u^{(i)}_\alpha, \varphi^{(i)}_3 \}$ $(i=1,2,3)$ and $(\varphi, \psi_1, \psi_2)$ are found.

Let us prove that the system (5.84) can always solved for a_1, a_2, a_3 and a_4. In the view of (5.9) and (5.77),

$$U(v) = \frac{1}{2} \sum_{i,j=1}^{4} \langle v^{(i)}, v^{(j)} \rangle a_i a_j \ .$$

Since $W(v)$ is positive definite and $v^{(i)}$ is not a rigid motion, it follows that

$$\det \langle v^{(i)}, v^{(j)} \rangle \neq 0 \quad (i,j=1,2,3,4) \ . \qquad (5.85)$$

By (5.11),(5.12),(5.78),(5.81) and $v^{(i)} \in \Lambda$ (i=1,2,3,4),

$$\langle v^{(\alpha)}, v^{(\beta)} \rangle = \int_{\partial B} [v_j^{(\alpha)} s_j(v^{(\beta)}) + \omega_j^{(\alpha)} q_j(v^{(\beta)})]\, da =$$

$$= -\frac{1}{2} h^2 \int_{\Sigma_2} t_{3\alpha}(v^{(\beta)})\, da + hD_{\alpha\beta} ,$$

$$\langle v^{(\alpha)}, v^{(3)} \rangle = h\, D_{\alpha 3} , \quad \langle v^{(3)}, v^{(3)} \rangle = h\, D_{33} ,$$

$$\langle v^{(i)}, v^{(4)} \rangle = 0 , \quad \langle v^{(4)}, v^{(4)} \rangle = h\, D .$$

Since $v^{(i)} \in \Lambda$ and $v^{(i)}_{,3} \in \mathcal{R}^*$, by Theorem 5.1 and (5.17) we find that $R_\alpha(v^{(i)}) = 0$. Thus,

$$\langle v^{(i)}, v^{(j)} \rangle = hD_{ij}, \quad \langle v^{(i)}, v^{(4)} \rangle = 0, \quad \langle v^{(4)}, v^{(4)} \rangle = hD . \qquad (5.86)$$

It follows from (5.12),(5.85) and (5.86) that $D_{ij} = D_{ji}$ and

$$\det(D_{ij}) \neq 0, \; D \neq 0 . \qquad (5.87)$$

We conclude that the system (5.83) uniquely determines the constants a_1, a_2, a_3 and a_4. □

Remark 1. The proof of this theorem offers a constructive procedure to obtain a solution of the extension-bending-torsion problem. This solution is given by (5.79) where the functions $u_\alpha^{(i)}, \varphi_3^{(i)}$ (i=1,2,3) are solutions of the plane strain problems (5.73),(5.74), the set of functions $(\varphi, \psi_1, \psi_2)$ is characterized by the boundary-value problem (5.69), and the constants a_1, a_2, a_3 and a_4 are determined by (5.83). Let us note that the torsion problem can be treated independently of the extension and bending problems.

Remark 2. The functions $u_\alpha^{(3)}$ and $\varphi_3^{(3)}$ can be determined in the following way. The corresponding equilibrium equations and boundary conditions are satisfied if one chooses

$$T_{\beta\alpha}(w^{(3)}) = -\lambda\delta_{\alpha\beta} \;, \quad M_{\alpha 3}(w^{(3)}) = 0.$$

Clearly, since λ is constant, the above functions satisfy the compatibility conditions [35]. It follows from the constitutive equations that

$$u^{(3)}_{1,1} = u^{(3)}_{2,2} = -\nu \;, \quad u^{(3)}_{1,2} + \varphi^{(3)}_3 = u^{(3)}_{2,1} - \varphi^{(3)}_3 = 0, \; \varphi^{(3)}_{,\alpha} = 0,$$

where $\nu = \lambda(2\lambda+2\mu+\varkappa)^{-1}$.

The integration of these equations yields

$$u^{(3)}_{\alpha} = -\nu x_{\alpha} \;, \quad \varphi^{(3)}_3 = 0,$$

modulo a plane rigid displacement.

It follows from (5.84) that

$$D_{\alpha 3} = D_{3\alpha} = AEx^{0}_{\alpha} \;, \; D_{33} = EA, \tag{5.88}$$

where A is the area of the cross-section, x^{0}_{α} are the coordinates of the centroid of Σ_1 and

$$E = (2\mu+\varkappa)(3\lambda+2\mu+\varkappa)/(2\lambda+2\mu+\varkappa).$$

Let us note that we established the relations $D_{3\alpha} = AEx^{0}_{\alpha}$ without recourse to the determination of $u^{(\rho)}_{\gamma}$. Another proof of this result can be obtained with the aid of the following formula which applies for the plane strain problem

$$\int_{\Gamma} x_{\alpha} s_{\alpha}(u)ds + \int_{\Sigma} x_{\alpha} f_{\alpha} da = \int_{\Sigma} [(x_{\alpha} t_{\beta\alpha}(u))_{,\beta} + x_{\alpha} f_{\alpha}]da =$$

$$= \int_{\Sigma} t_{\beta\beta}(u)da = (2\lambda+2\mu+\varkappa)\int_{\Sigma} u_{\rho,\rho}\, da.$$

Remark 3. If the rectangular Cartesian coordinate frame is chosen in such a way that the origin O coincides with the centroid of the cross-section Σ_1, then the problems of extension and bending can be treated independently one of the other. This aspect does not appear in the solution presented in [59].

In view of (5.79),(5.83) and (5.88), if $x^0_\alpha = 0$ then we find the following solutions:

(i) Extension solution ($F_\alpha = 0$, $M_i = 0$):

$$u_\alpha = -a_3 \nu x_\alpha , \quad u_3 = a_3 x_3, \quad \varphi_i = 0,$$

where

$$EAa_3 = -F_3.$$

(ii) Bending solution ($F_i = 0$, $M_3 = 0$):

$$u_\alpha = -\frac{1}{2} a_\alpha x_3^2 + \sum_{\rho=1}^{2} a_\rho u_\alpha^{(\rho)} , \quad u_3 = a_\beta x_\beta x_3 ,$$

$$\varphi_\alpha = \varepsilon_{\alpha\beta} a_\beta x_3 , \qquad \varphi_3 = \sum_{\rho=1}^{2} a_\rho \varphi_3^{(\rho)} , \tag{5.89}$$

where the functions $u_\alpha^{(\rho)}, \varphi_3^{(\rho)}$ ($\rho = 1,2$) are solutions of the corresponding plane strain problems from (5.73),(5.74), and the constants a_1, a_2 are determined by

$$D_{\alpha\beta} a_\beta = \varepsilon_{\alpha\eta} M_\eta .$$

(iii) Torsion solution ($F_i = 0$, $M_\alpha = 0$):

$$u_\alpha = \varepsilon_{\beta\alpha} a_4 x_\beta x_3 , \quad u_3 = a_4 \varphi ,$$

$$\varphi_\alpha = a_4 \psi_\alpha , \quad \varphi_3 = a_4 x_3 , \tag{5.90}$$

where the torsion functions φ, ψ_1 and ψ_2 are characterized by the boundary-value problem (5.69) and a_4 is given by

$$Da_4 = -M_3 .$$

D is the torsional rigidity for micropolar cylinders.

5.7. Flexure

By a solution of flexure problem we mean any vector field $u \in \Lambda$ that satisfies the conditions

$$R_\alpha(u) = F_\alpha , \quad R_3(u) = 0 , \quad H_i(u) = 0 . \tag{5.91}$$

Let $\hat{a} = (a_1, a_2, a_3, a_4)$. We denote for the remainder of this chapter by $v\{\hat{a}\}$ the vector field v defined by (5.79).

In view of Corollaries 5.1, 5.2 and Theorem 5.7, we are led to seek a solution of the flexure problem in the form

$$u = \int_0^{x_3} v\{\hat{b}\}dx_3 + v\{\hat{c}\} + w' , \tag{5.92}$$

where $\hat{b} = (b_1, b_2, b_3, b_4)$ and $\hat{c} = (c_1, c_2, c_3, c_4)$ are two constant four-dimensional vectors, and $w' = (w_i', \chi_i')$ is a vector field independent of x_3 such that $w' \in C^1(\bar{\Sigma}) \cap C^2(\Sigma)$.

Theorem 5.8. Let Y be the set of all vector fields of the form (5.92). Then there exists a vector field $u^0 \in Y$ which is solution of the problem (P_2).

Proof. Let $u^0 \in Y$. Let us prove that the vector field $w' = (w_i', \chi_i')$ and the constants b_i, c_i (i=1,2,3,4) can be determined so that $u^0 \in K_{II}(F_1, F_2)$. First we determine the vector $\hat{b}$. Thus, if $u^0 \in K_{II}(F_1, F_2)$ then by Corollary 5.2 and (5.92),

$$v\{\hat{b}\} \in K_I(0, F_2, -F_1, 0). \tag{5.93}$$

In view of (5.83) and (5.93) we obtain

$$D_{\alpha j} b_j = -F_\alpha , \quad D_{3j} b_j = 0 , \quad b_4 = 0 . \tag{5.94}$$

This system can always be solved for b_1, b_2 and b_3.

By (5.79),(5.92) and (5.94), we find that

$$
\begin{aligned}
u_\alpha^o &= -\tfrac{1}{6}b_\alpha x_3^3 - \tfrac{1}{2}c_\alpha x_3^2 - c_4\varepsilon_{\alpha\beta}x_\beta x_3 + \sum_{i=1}^{3}(b_i x_3 + c_i)u_\alpha^{(i)} + w'_\alpha , \\
u_3^o &= \tfrac{1}{2}(b_\rho x_\rho + b_3)x_3^2 + (c_\rho x_\rho + c_3)x_3 + c_4\varphi + w'_3 , \\
\varphi_\alpha^o &= \tfrac{1}{2}\varepsilon_{\alpha\beta}b_\beta x_3^2 + \varepsilon_{\alpha\beta}c_\beta x_3 + c_4\psi_\alpha + \chi'_\alpha , \\
\varphi_3^o &= c_4 x_3 + \sum_{i=1}^{3}(b_i x_3 + c_i)\varphi_3^{(i)} + \chi'_3 ,
\end{aligned}
\tag{5.95}
$$

where $(u_\alpha^{(i)}, \varphi_3^{(i)})$ (i=1,2,3) are characterized by (5.73) and (5.74). It follows from (5.1),(5.5) and (5.95) that

$$
\begin{aligned}
t_{\alpha\beta}(u^o) &= \lambda[(b_\rho x_\rho + b_3)x_3 + c_\rho x_\rho + c_3]\delta_{\alpha\beta} + \sum_{i=1}^{3}(b_i x_3 + c_i)T_{\alpha\beta}(w^{(i)}) + \\
&\quad + T_{\alpha\beta}(\omega^o), \\
t_{33}(u^o) &= (\lambda + 2\mu + \varkappa)[(b_\rho x_\rho + b_3)x_3 + c_\rho x_\rho + c_3] + \\
&\quad + \lambda\sum_{i=1}^{3}(b_i x_3 + c_i)u_{\rho,\rho}^{(i)} + \lambda w'_{\rho,\rho} , \\
t_{\alpha 3}(u^o) &= P_\alpha(\overline{w}) + c_4[P_\alpha(\hat{w}) + \mu\varepsilon_{\beta\alpha}x_\beta] + \mu\sum_{i=1}^{3} b_i u_\alpha^{(i)} , \\
t_{3\alpha}(u^o) &= Q_\alpha(\overline{w}) + c_4[Q_\alpha(\hat{w}) + (\mu + \varkappa)\varepsilon_{\beta\alpha}x_\beta] + (\mu + \varkappa)\sum_{i=1}^{3} b_i u_\alpha^{(i)} , \\
m_{\lambda\nu}(u^o) &= H_{\lambda\nu}(\overline{w}) + c_4[\alpha\delta_{\lambda\nu} + H_{\lambda\nu}(\hat{w})] + \alpha\delta_{\lambda\nu}\sum_{i=1}^{3} b_i\varphi_3^{(i)} , \\
m_{33}(u^o) &= \alpha(c_4\psi_{\rho,\rho} + \chi'_{\rho,\rho}) + (\alpha + \beta + \gamma)(c_4 + \sum_{i=1}^{3} b_i\varphi_3^{(i)}),
\end{aligned}
\tag{5.96}
$$

$$m_{\alpha 3}(u^0)=\beta\varepsilon_{\alpha\nu}(b_\nu x_3+c_\nu)+\sum_{i=1}^{3}(b_i x_3+c_i)M_{\alpha 3}(w^{(i)})+M_{\alpha 3}(\omega^0),$$

$$m_{3\alpha}(u^0)=\gamma\varepsilon_{\alpha\nu}(b_\nu x_3+c_\nu)+\beta\sum_{i=1}^{3}(b_i x_3+c_i)\varphi_{3,\alpha}^{(i)}+\beta\chi'_{3,\alpha},$$

where we have used the notations

$$\omega^0=(w'_1,w'_2,0,0,0,\chi'_3),\quad \bar{w}=(0,0,w'_3,\chi'_1,\chi'_2,0).$$

Substituting (5.96) into equations of equilibrium we find, with the aid of (5.69) and (5.73), that

$$(T_{\beta\alpha}(\omega^0))_{,\beta}=0,\ (M_{\rho 3}(\omega^0))_{,\rho}+\varepsilon_{\alpha\beta}T_{\alpha\beta}(\omega^0)=0 \text{ on } \Sigma \quad (5.97)$$

and

$$L_i\,\bar{w}=\xi_i \quad \text{on } \Sigma, \quad (5.98)$$

where

$$\xi_\nu=-\gamma\varepsilon_{\nu\rho}b_\rho-\sum_{i=1}^{3}b_i[(\alpha+\beta)\varphi_{3,\nu}^{(i)}-\varepsilon_{\nu\beta}\varkappa u_\beta^{(i)}],$$

$$\xi_3=-(\lambda+2\mu+\varkappa)(b_\rho x_\rho+b_3)-(\lambda+\mu)\sum_{i=1}^{3}b_i u_{\rho,\rho}^{(i)}.$$

By (5.69) and (5.74), the conditions on the lateral boundary reduce to

$$T_{\beta\alpha}(\omega^0)n_\beta=0,\quad M_{\alpha 3}(\omega^0)n_\alpha=0 \quad \text{on } \Gamma, \quad (5.99)$$

and

$$N_\rho\bar{w}=-\alpha n_\rho\sum_{i=1}^{3}b_i\varphi_3^{(i)},\quad N_3\bar{w}=-\mu n_\alpha\sum_{i=1}^{3}b_i u_\alpha^{(i)} \text{ on } \Gamma. \quad (5.100)$$

The relations (5.97) and (5.99) constitute a plane strain problem corresponding to null data. We conclude that $w'_\alpha=0$, $\chi'_3=0$ (modulo a plane rigid motion).

The necessary and sufficient condition for the existence of a solution to the boundary-value problem (5.98) and (5.100) reduces to

$$D_{3i}\, b_i = 0 .$$

This condition is satisfied on the basis of (5.94). Thus, the functions w_3' , χ_α' are characterized by the boundary-value problem (5.98), (5.100).

Clearly, the conditions $R_\alpha(u) = F_\alpha$ are satisfied by (5.93),(5.94) and Theorem 5.1. The conditions $R_3(u) = 0$, $\underline{H}(u) = \underline{0}$ reduce to

$$D_{ij}\, c_j = 0, \tag{5.101}$$

and

$$\begin{aligned} Dc_4 = - \int_\Sigma \Big\{ & \varepsilon_{\alpha\beta} x_\alpha \Big[\mu w_{3,\beta}' + \varepsilon_{\nu\beta} \varkappa \chi_\nu' + (\mu + \varkappa) \sum_{i=1}^{3} b_i u_\beta^{(i)}\Big] + \\ & + (\alpha + \beta + \gamma) \sum_{i=1}^{3} b_i \varphi_3^{(i)} + \alpha \chi_{\rho,\rho}' \Big\} da . \end{aligned} \tag{5.102}$$

By (5.87) and (5.101) we conclude that $c_i = 0$. The constant c_4 is given by (5.102). □

<u>Remark 1.</u> The proof of the above theorem offers a constructive procedure to obtain a solution of the flexure problem. This solution is given by (5.95) where $w_\alpha' = \chi_3' = 0$, $c_i = 0$, the functions $u_\alpha^{(i)}$, $\varphi_3^{(i)}$ $(i=1,2,3)$ are solutions of the plane strain problems (5.73),(5.74), the functions φ, ψ_α are characterized by the boundary value problem (5.69), the functions w_3', χ_α' are characterized by the boundary-value problem (5.98),(5.100), and the constants b_i and c_4 are determined by (5.94) and (5.102).

<u>Remark 2.</u> If the rectangular Cartesian coordinate frame is chosen in such a way that the origin O coincides with the centroid of the cross-section Σ_1, then (5.88) implies $D_{3\alpha} = 0$. It follows from (5.94) that $b_3 = 0$. In this case, (5.95) yields the following solution of the flexure problem

$$\begin{aligned} u_\alpha^0 &= -\frac{1}{6} b_\alpha x_3^3 + c_4 \varepsilon_{\beta\alpha} x_\beta x_3 + x_3 \sum_{\rho=1}^{2} b_\rho u_\alpha^{(\rho)}, \\ u_3^0 &= \frac{1}{2} (b_1 x_1 + b_2 x_2) x_3^2 + c_4 \varphi + w_3' , \end{aligned} \tag{5.103}$$

$$\varphi_\alpha^0 = \frac{1}{2}\varepsilon_{\alpha\beta}b_\beta x_3^2 + c_4\psi_\alpha + \chi'_\alpha ,$$

$$\varphi_3^0 = c_4x_3 + x_3\sum_{\rho=1}^{2} b_\rho\varphi_3^{(\rho)} ,$$

where the constants b_α are determined by

$$D_{\alpha\beta}b_\beta = -F_\alpha ,$$

and c_4 is given by (5.102).

The stress tensor and couple stress tensor are given by

$$t_{\alpha\beta}(u^0) = \lambda b_\rho x_\rho x_3\delta_{\alpha\beta} + x_3\sum_{\rho=1}^{2} b_\rho T_{\alpha\beta}(w^{(\rho)}),$$

$$t_{33}(u^0) = (\lambda+2\mu+\varkappa)b_\rho x_\rho x_3 + x_3\lambda\sum_{\rho=1}^{2} b_\rho u_{\nu,\nu}^{(\rho)},$$

$$t_{\alpha3}(u^0) = P_\alpha(\bar{w})+c_4[P_\alpha(\hat{w})+\mu\varepsilon_{\beta\alpha}x_\beta]+\mu\sum_{\rho=1}^{2} b_\rho u_\alpha^{(\rho)},$$

$$t_{3\alpha}(u^0) = Q_\alpha(\bar{w})+c_4[Q_\alpha(\hat{w})+\varepsilon_{\beta\alpha}(\mu+\varkappa)x_\beta]+(\mu+\varkappa)\sum_{\rho=1}^{2} b_\rho u_\alpha^{(\rho)},$$

$$m_{\lambda\nu}(u^0) = H_{\lambda\nu}(\bar{w})+c_4[H_{\lambda\nu}(\hat{w})+\alpha\delta_{\lambda\nu}]+\alpha\delta_{\lambda\nu}\sum_{\rho=1}^{2} b_\rho\varphi_3^{(\rho)},$$

$$m_{33}(u^0) = \alpha(c_4\psi_{\rho,\rho}+\chi'_{\rho,\rho})+(\alpha+\beta+\gamma)(c_4+\sum_{\rho=1}^{2} b_\rho\varphi_3^{(\rho)}),$$

$$m_{\alpha3}(u^0) = \beta\varepsilon_{\alpha\nu}x_3b_\nu + x_3\sum_{\rho=1}^{2} b_\rho M_{\alpha3}(w^{(\rho)}),$$

$$m_{3\alpha}(u^0) = \gamma\varepsilon_{\alpha\nu}b_\nu x_3 + \beta x_3\sum_{\rho=1}^{2} b_\rho\varphi_{3,\alpha}^{(\rho)} .$$

<u>Remark 3.</u> It is a simple matter to see that if we replace (5.101) and (5.102) by

$$D_{\alpha j}c_j = \varepsilon_{\alpha\rho}M_\rho , \quad D_{3j}c_j = -F_3,$$

$$Dc_4 = -M_3 - \int_{\Sigma}\{\varepsilon_{\alpha\beta}x_\alpha[\mu w'_{3,\beta} + \varepsilon_{\nu\beta}\varkappa\chi'_\nu + (\mu+\varkappa)\sum_{i=1}^{3} b_i u_\beta^{(i)} + (\alpha+\beta+\gamma)\sum_{i=1}^{3} b_i\varphi_3^{(i)} + \alpha\chi'_{\rho,\rho}\}da ,$$

then the vector field u^0 defined by (5.95) belongs to $K(\underline{F},\underline{M})$.

5.8. Minimum Principles

In this section we present minimum strain-energy characterizations of the solutions obtained in Sections 5.6 and 5.7. We assume that the origin O coincides with the centroid of Σ_1. Since in the extension solution the microrotation vector vanishes, we renounce to study this solution.

Let A_I denote the set of all equilibrium vector fields u that satisfy the conditions

$$\underline{s}(u) = \underline{0},\ \underline{q}(u) = \underline{0} \text{ on } \Pi,\ t_{3\rho}(u) = 0,\ m_{33}(u) = 0 \text{ on } \Sigma_\beta, \qquad H_\alpha(u) = M . \tag{5.104}$$

Theorem 5.9. Let v be the solution (5.89) of the bending problem, corresponding to a couple of moment $\underline{M}(M_1,M_2,0)$. Then

$$U(v) \leq U(u) ,$$

for every $u \in A_I$, and equality holds only if u = v (modulo a rigid motion).

Proof. Note that $v = (v_i,\omega_i) \in A_I$. Let $u \in A_I$ and define $u' = u - v$. Then u' is an equilibrium vector field that satisfies

$$\underline{t}(u')=\underline{0},\ \underline{q}(u')=\underline{0} \text{ on } \Pi,\ t_{3\rho}(u')=0,\ m_{33}(u')=0 \text{ on } \Sigma_\beta, \qquad H_\alpha(u')=0 . \tag{5.105}$$

Clearly,

$$U(u) = U(u') + U(v) + \langle u',v\rangle \; . \tag{5.106}$$

It follows from (5.11),(5.17),(5.89) and (5.105) that

$$\begin{aligned}\langle u',v\rangle &= \int_{\partial B} [v_i s_i(u') + \omega_i q_i(u')]da = \\ &= \int_{\Sigma_2} [v_i t_{3i}(u') + \omega_i m_{3i}(u')]da - \\ &- \int_{\Sigma_1} [v_i t_{3i}(u') + \omega_i m_{3i}(u')]da = \\ &= h\int_{\Sigma_2} [(a_1x_1 + a_2x_2)t_{33}(u') + \varepsilon_{\alpha\beta}a_\beta m_{3\alpha}(u')]da = \\ &= ha_\alpha[\varepsilon_{\alpha\beta}H_\beta(u') - hR_\alpha(u')] = 0 \; .\end{aligned} \tag{5.107}$$

We conclude from (5.106) and (5.107) that $U(u) \geq U(v)$, and $U(u) = U(v)$ only if u' is a rigid motion. □

Let A_{II} denote the set of all equilibrium vector fields u that satisfy the conditions

$$\underline{s}(u)=\underline{0},\; \underline{q}(u)=\underline{0} \text{ on } \Pi,\; t_{33}(u)=0,\; m_{3\alpha}(u)=0 \text{ on } \Sigma_\rho,\; H_3(u)=M_3 \,. \tag{5.108}$$

<u>Theorem 5.10.</u> Let v be the solution (5.90) of the torsion problem corresponding to the scalar torque M_3. Then

$$U(v) \leq U(u) \;,$$

for every $u \in A_{II}$, and equality holds only if u = v.

<u>Proof.</u> Let $u \in A_{II}$ and $v = (v_i, \omega_i)$. Since $v \in A_{II}$ it follows that the field $u' = u - v$, is an equilibrium vector field that satisfies

$$\underline{s}(u') = \underline{0},\ \underline{q}(u') = \underline{0} \text{ on } \Pi,\ t_{33}(u') = 0, \qquad (5.109)$$
$$m_{3\alpha}(u') = 0 \text{ on } \Sigma_\rho,\ H_3(u') = 0.$$

In view of (5.11),(5.17),(5.90) and (5.109) we obtain

$$\langle u',v\rangle = a_4 h\int_{\Sigma_2}[\varepsilon_{\beta\alpha}x_\beta t_{3\alpha}(u') + m_{33}(u')]da = -a_4 hH_3(u') = 0.$$

This result implies that

$$U(u) - U(v) = U(u-v).$$

The conclusion is now immediate. □

Let E denote the set of all equilibrium vector fields u that satisfy the conditions

$$u_{,3}\in C^1(\bar{B})\cap C^2(B),\quad \underline{s}(u) = \underline{0},\quad \underline{q}(u) = \underline{0} \text{ on } \Pi,$$
$$[t_{3\alpha}(u_{,3})](x_1,x_2,0) = [t_{3\alpha}(u_{,3})](x_1,x_2,h),$$
$$[m_{33}(u_{,3})](x_1,x_2,0)=[m_{33}(u_{,3})](x_1,x_2,h)\ \ (x_1,x_2)\in\Sigma,$$
$$R_\alpha(u) = F_\alpha.$$

<u>Theorem 5.11.</u> Let u^0 be the solution (5.103) of the flexure problem corresponding to the loads F_1 and F_2. Then

$$U(u^0_{,3}) \leq U(u_{,3}),$$

for every $u\in E$, and equality holds only if $u_{,3} = u^0_{,3}$.

<u>Proof.</u> Let $u\in E$. Since $u^0\in E$ it follows that the vector field u' defined by $u' = u-u^0$ is an equilibrium displacement field that satisfies

$$u'_{,3}\in C^1(\bar{B})\cap C^2(B),\ \underline{s}(u') = \underline{0},\ \underline{q}(u') = \underline{0} \text{ on } \Pi,$$
$$[t_{3\alpha}(u'_{,3})](x_1,x_2,0) = [t_{3\alpha}(u'_{,3})](x_1,x_2,h), \qquad (5.110)$$
$$[m_{33}(u'_{,3})](x_1,x_2,0)=[m_{33}(u'_{,3})](x_1,x_2,h)\ \ (x_1,x_2)\in\Sigma,\ R_\alpha(u')=0.$$

Clearly,

$$U(u_{,3}) = U(u'_{,3}) + U(u^{0}_{,3}) + \langle u'_{,3} , u^{0}_{,3} \rangle .$$

By (5.11),(5.17),(5.103) and (5.110),

$$\langle u'_{,3} , u^{0}_{,3} \rangle = \int_{\partial B} [u^{0}_{i,3} s_i(u'_{,3}) + \varphi^{0}_{i,3} q_i(u'_{,3})] da =$$

$$= -\frac{1}{2} b_\alpha h^2 \int_{\Sigma_2} t_{3\alpha}(u'_{,3}) da +$$

$$+ hb_\alpha \int_{\Sigma_2} [x_\alpha t_{33}(u'_{,3}) + \varepsilon_{\beta\alpha} m_{3\beta}(u'_{,3})] da =$$

$$= \frac{1}{2} b_\alpha h^2 R_\alpha(u'_{,3}) + hb_\alpha(\varepsilon_{\alpha\beta} H_\beta(u'_{,3}) - hR_\alpha(u'_{,3})) =$$

$$= -\frac{1}{2} b_\alpha h^2 R_\alpha(u'_{,3}) + hb_\alpha \varepsilon_{\alpha\beta} H_\beta(u'_{,3}).$$

In view of Theorem 5.1 and (5.110) we conclude that

$$\langle u'_{,3} , u^{0}_{,3} \rangle = 0 .$$

Thus

$$U(u_{,3}) = U(u'_{,3}) + U(u^{0}_{,3}).$$

The desired conclusion is now immediate. □

5.9. Global Strain Measures

In this section we study Truesdell's problem for Cosserat elastic cylinders. Following [123], we are led to a global measure of strain appropriate to torsion. Truesdell's problem rephrased for the flexure is also considered.

We first consider Truesdell's problem for the torsion of Cosserat elastic cylinders. We denote by T the set of all solutions of the torsion problem corresponding to the scalar torque M_3. We are led to the following problem: to define the functional $\tau(\cdot)$ on T such that

$$M_3 = D\tau(u) \quad \text{for every} \quad u \in T . \tag{5.111}$$

We denote by T_0 the set of all equilibrium vector fields u that satisfy the conditions

$$\underline{s}(u) = \underline{0} \, , \; \underline{q}(u) = \underline{0} \text{ on } \Pi \, , \; t_{33}(u) = 0, \; m_{3\alpha}(u) = 0 \text{ on } \Sigma_\rho \, ,$$
$$R_\alpha(u) = 0 \, , \; H_3(u) = M_3 \, . \tag{5.112}$$

Clearly, if $u \in T_0$, then $R_3(u) = 0$, $H_\alpha(u) = 0$, so that $u \in T$. We define the real function

$$\xi \longrightarrow \| u - \xi v^{(4)} \|_e^2 \, ,$$

where $u \in T_0$ and $v^{(4)}$ is given by (5.78). This function attains its minimum at

$$\gamma(u) = \langle u, v^{(4)} \rangle / \| v^{(4)} \|_e^2 \, . \tag{5.113}$$

Let us prove that $\gamma(u) = \tau(u)$ for every $u \in T_0$. In view of (5.11), (5.78) and (5.112) we find that

$$\langle u, v^{(4)} \rangle = h \, H_3(u) \, . \tag{5.114}$$

With the aid of the relations (5.81) we obtain

$$\| v^{(4)} \|_e^2 = h \, D \, , \tag{5.115}$$

where D is defined in (5.84). Thus, from (5.113), (5.114) and (5.115) we arrive at

$$H_3(u) = D \, \gamma(u) \, . \tag{5.116}$$

It follows from (5.111) and (5.116) that $\tau(u) = \gamma(u)$ for each $u \in T_0$. On the other hand, by (5.11), (5.12) and (5.78) we find that

$$\langle u, v^{(4)} \rangle = N(u), \tag{5.117}$$

where

$$N(u) = \int_{\Sigma_2} \{u_\alpha[\mu\varphi_{,\alpha} + \varkappa\varepsilon_{\beta\alpha}\psi_\beta + \tfrac{1}{2}\varepsilon_{\beta\alpha}(2\mu+\varkappa)x_\beta] + \varphi_3(\alpha\psi_{\rho,\rho} + \beta + \gamma)\}da - \int_{\Sigma_1}\{u_\alpha[\mu\varphi_{,\alpha} + \varkappa\varepsilon_{\beta\alpha}\psi_\beta + \tfrac{1}{2}\varepsilon_{\beta\alpha}(2\mu+\varkappa)x_\beta] + \varphi_3(\alpha\psi_{\rho,\rho}+\beta+\gamma)\}da .$$

Thus, from (5.113),(5.115) and (5.117) we conclude that

$$\tau(u) = \frac{1}{hD} N(u) \quad \text{for each } u \in T_0 .$$

This relation defines the generalized twist on the subclass T_0 of solutions to the torsion problem. In view of (5.111), we interpret the right-hand side of the above relation as the global measure of strain appropriate to torsion.

We assume for the remainder of this section that the rectangular Cartesian coordinate is chosen in such a way that the origin O coincides with the centroid of the cross-section Σ_1.

Truesdell's problem can be set also for the flexure. Thus we are led to the following problem: to define the functionals $\eta_\alpha(\cdot)$ on $K_{II}(F_1,F_2)$ such that

$$D_{\alpha\rho}\eta_\rho(u) = - F_\alpha , \tag{5.118}$$

for each $u \in K_{II}(F_1,F_2)$.

Let G denote the set of all equilibrium vector fields u that satisfy the conditions

$$\begin{gathered} u_{,3} \in C^1(\bar{B}) \cap C^2(B), \quad \underline{s}(u) = \underline{0}, \; \underline{q}(u) = \underline{0} \text{ on } \Pi, \\ [t_{3\alpha}(u_{,3})](x_1,x_2,0) = [t_{3\alpha}(u_{,3})](x_1,x_2,h), \\ [m_{33}(u_{,3})](x_1,x_2,0) = [m_{33}(u_{,3})](x_1,x_2,h) , \quad (x_1,x_2) \in \Sigma , \\ R_\alpha(u) = F_\alpha , \quad R_3(u) = 0 , \quad \underline{H}(u) = \underline{0} . \end{gathered} \tag{5.119}$$

Clearly, if $u \in G$, then $u \in K_{II}(F_1,F_2)$. Let us consider the real function f defined by

$$f(\xi_1,\xi_2) = 2U(u_{,3} - \xi_1 v^{(1)} - \xi_2 v^{(2)}), \qquad (5.120)$$

where $u\in G$ and $v^{(\rho)}$ $(\rho = 1,2)$ are given by (5.78). By (5.86) and (5.120),

$$f = hD_{\alpha\beta}\xi_\alpha\xi_\beta - 2\xi_\alpha\langle u_{,3}\ ,\ v^{(\alpha)}\rangle + \langle u_{,3}\ ,\ u_{,3}\rangle\ .$$

Since $D_{\alpha\beta}$ is positive definite, f will be a minimum at $(\rho_1(u), \rho_2(u))$ if and only if $(\rho_1(u), \rho_2(u))$ is the solution of the following system of equations

$$h\, D_{\alpha\beta}\rho_\beta(u) = \langle u_{,3}\ ,\ v^{(\alpha)}\rangle\ . \qquad (5.121)$$

Let us prove that $\rho_\alpha(u) = \eta_\alpha(u)$ $(\alpha = 1,2)$, for every $u\in G$. By (5.11), (5.17),(5.78) and (5.119) we obtain

$$\langle u_{,3}\ ,v^{(\alpha)}\rangle = \int_{\partial B}[v_i^{(\alpha)}s_i(u_{,3}) + \omega_i^{(\alpha)}q_i(u_{,3})]da =$$
$$= -\frac{1}{2}h^2 R_\alpha(u_{,3}) + h\,\varepsilon_{\alpha\beta}H_\beta(u_{,3}). \qquad (5.122)$$

In view of (5.122) and Theorem 5.1 we find that

$$\langle u_{,3}\ ,\ v^{(\alpha)}\rangle = -h\, R_\alpha(u). \qquad (5.123)$$

It follows from (5.121) and (5.123) that

$$D_{\alpha\beta}\rho_\beta(u) = -R_\alpha(u)\ . \qquad (5.124)$$

Thus, from (5.118),(5.119) and (5.124) we conclude that $\eta_\alpha(u) = \rho_\alpha(u)$ $(\alpha = 1,2)$ for each $u\in G$.

On the other hand, by (5.11),(5.12) and (5.81) we find

$$\langle u_{,3},v^{(\alpha)}\rangle = \int_{\partial B}[u_{i,3}t_{3i}(v^{(\alpha)}) + \varphi_{i,3}m_{3i}(v^{(\alpha)})]da = S_\alpha(u), \qquad (5.125)$$

where

$$S_\rho(u) = \int_{\Sigma_2} \{u_{3,3}[(\lambda+2\mu+\varkappa)x_\rho + \lambda u^{(\rho)}_{\nu,\nu}] + \varphi_{\alpha,3}[\gamma\varepsilon_{\alpha\rho} + \beta\varphi^{(\rho)}_{3,\alpha}]\}da -$$

$$- \int_{\Sigma_1} \{u_{3,3}[(\lambda+2\mu+\varkappa)x_\rho + \lambda u^{(\rho)}_{\nu,\nu}] + \varphi_{\alpha,3}[\gamma\varepsilon_{\alpha\rho} + \beta\varphi^{(\rho)}_{3,\alpha}]\}da ,$$

for each $u = (u_i, \varphi_i) \in G$.

By (5.121) and (5.125) we obtain

$$D_{\alpha\beta}\eta_\beta(u) = \frac{1}{h} S_\alpha(u) \quad (\alpha = 1,2),$$

for every $u \in G$. This system defines $\eta_\alpha(\cdot)$ on the subclass G of solutions to the flexure problem. Following [123], we can interpret $\eta_\alpha(u)$ as the global measures of strain appropriate to flexure associated with $u \in G$.

Truesdell's problem can be set and solved also for extension and for bending.

5.10. Theory of Loaded Cylinders.

We assume that the body force $\underline{f}$ and the body couple $\underline{g}$ are prescribed on B. By an equilibrium vector field on B corresponding to the body loads $\{\underline{f}, \underline{g}\}$ we mean a six-dimensional vector field $u \in C^1(\bar{B}) \cap C^2(B)$ that satisfies the equations

$$(t_{ji}(u))_{,j} + f_i = 0, \quad (m_{ji}(u))_{,j} + \varepsilon_{ijk}t_{jk}(u) + g_i = 0, \qquad (5.126)$$

on B. We assume now that the conditions (5.15) are replaced by

$$\underline{s}(u) = \underline{p}, \quad \underline{q}(u) = \underline{k} \text{ on } \Pi, \quad \underline{R}(u) = \underline{F}, \quad \underline{H}(u) = \underline{M}, \qquad (5.127)$$

where $\underline{p}$ and $\underline{k}$ are prescribed vector fields, and $\underline{F}$ and $\underline{M}$ are prescribed vectors. The problem of loaded cylinder consists in finding an equilibrium vector field on B that corresponds to the body loads $\{\underline{f}, \underline{g}\}$ and satisfies the conditions (5.127).

When $\underline{f}, \underline{g}, \underline{p}$ and $\underline{k}$ are independent of the axial coordinate we refer to this problem as Almansi-Michell problem. We continue to denote by (P_3) the Almansi-Michell problem corresponding to the system of loads

$\{\underline{F},\underline{M},\underline{f},\underline{g},\underline{p},\underline{k}\}$. Let $K_{III}(\underline{F},\underline{M},\underline{f},\underline{g},\underline{p},\underline{k})$ denote the class of solutions to the problem (P_3).

The next theorem will be used in the following.

<u>Theorem 5.12.</u> If $u \in C^1(\bar{B}) \cap C^2(B)$, then

$$R_i(u_{,3}) = \int_{\partial\Sigma_1} s_i(u)ds - \int_{\Sigma_1} (t_{ji}(u))_{,j} da,$$

$$H_\alpha(u_{,3}) = \int_{\partial\Sigma_1} [\varepsilon_{\alpha\beta}x_\beta s_3(u) + q_\alpha(u)]ds - \int_{\Sigma_1} [\varepsilon_{\alpha\beta}x_\beta (t_{j3}(u))_{,j} + (m_{j\alpha}(u))_{,j} + \varepsilon_{\alpha rs}t_{rs}(u)]da + \varepsilon_{\alpha\beta}R_\beta(u),$$

$$H_3(u_{,3}) = \int_{\partial\Sigma_1} [\varepsilon_{\alpha\beta}x_\alpha s_\beta(u) + q_3(u)]ds - \int_{\Sigma_1} [\varepsilon_{\alpha\beta}x_\alpha (t_{j\beta}(u))_{,j} + (m_{j3}(u))_{,j} + \varepsilon_{\alpha\beta}t_{\alpha\beta}(u)]da .$$

The proof of this theorem is analogous to that given for Theorem 5.1.

We consider first the problem (P_3). The Theorem 5.12 has the following immediate consequence

<u>Corollary 5.3.</u> If $u \in K_{III}(\underline{F},\underline{M},\underline{f},\underline{g},\underline{p},\underline{k})$ and $u_{,3} \in C^1(\bar{B}) \cap C^2(B)$, then $u_{,3} \in K(\underline{G},\underline{Z})$ where

$$\underline{G} = \int_\Gamma \underline{p}\, ds + \int_\Sigma \underline{f}\, da ,$$

$$Z_\alpha = \int_\Gamma (\varepsilon_{\alpha\beta}x_\beta p_3 + k_\alpha)ds + \int_\Sigma (\varepsilon_{\alpha\beta}x_\beta f_3 + g_\alpha)da + \varepsilon_{\alpha\beta}F_\beta , \tag{5.128}$$

$$Z_3 = \int_\Gamma (\varepsilon_{\alpha\beta}x_\alpha p_\beta + k_3)ds + \int_\Sigma (\varepsilon_{\alpha\beta}x_\alpha f_\beta + g_3)da .$$

In view of Corollary 5.3 and (5.92) we are led to seek a solution of the problem (P_3) in the form

$$u = \int_0^{x_3}\int_0^{x_3} v\{\hat{b}\}dx_3 dx_3 + \int_0^{x_3} v\{\hat{c}\}dx_3 + v\{\hat{d}\} + x_3 u' + u^0, \qquad (5.129)$$

where $\hat{b}$, $\hat{c}$ and $\hat{d}$ are unknown constant vectors, u' and u^0 are unknown vector fields independent of x_3, and $v\{\hat{a}\}$ is defined by (5.79).

Theorem 5.13. Let V be the set of all vector fields of the form (5.129). Then there exists a vector field $\hat{u} \in V$ which is solution of the problem (P_3).

Proof. We have to determine $\hat{b}, \hat{c}, \hat{d}, u'$ and u^0 such that $\hat{u} \in K_{III}(\underline{F},\underline{M},\underline{f},\underline{g},\underline{p},\underline{k})$. If $\hat{u} \in K_{III}(\underline{F},\underline{M},\underline{f},\underline{g},\underline{p},\underline{k})$, then by Corollaries 5.1, 5.3 and (5.129),

$$\int_0^{x_3} v\{\hat{b}\}dx_3 + v\{\hat{c}\} + u' \in K(\underline{G},\underline{Z}) .$$

In view of Theorem 5.8 and (5.94), it follows that

$$D_{\alpha j}\, b_j = - G_\alpha , \quad D_{3j}\, b_j = 0 , \quad b_4 = 0 , \qquad (5.130)$$

and $u' = (0,0,\chi,\chi_1,\chi_2,0)$ is characterized by

$$L_\nu u' = -\gamma \varepsilon_{\nu\rho} b_\rho - \sum_{i=1}^{3} b_i[(\alpha+\beta)\varphi^{(i)}_{3,\nu} - \varkappa\varepsilon_{\nu\beta}u^{(i)}_\beta],$$

$$L_3 u' = -(\lambda+2\mu+\varkappa)(b_\rho x_\rho + b_3) - (\lambda+\mu)\sum_{i=1}^{3} b_i u^{(i)}_{\rho,\rho} \quad \text{on } \Sigma, \qquad (5.131)$$

$$N_\rho u' = -\alpha n_\rho \sum_{i=1}^{3} b_i \varphi^{(i)}_3 , \quad N_3 u' = -\mu n_\alpha \sum_{i=1}^{3} b_i u^{(i)}_\alpha \quad \text{on } \Gamma .$$

Moreover, $\hat{c}$ is determined by

$$D_{\alpha j}\, c_j = \varepsilon_{\alpha\rho} G_\rho\ , \quad D_{3j}\, c_j = -G_3\ ,$$

$$Dc_4 = -Z_3 - \int_\Sigma \{\varepsilon_{\alpha\beta} x_\alpha [\mu \chi_{,\beta} + \varkappa \varepsilon_{\nu\beta} \chi_\nu + (\mu+\varkappa)\sum_{i=1}^{3} b_i u_\beta^{(i)}] + \qquad (5.132)$$

$$+ (\alpha+\beta+\gamma)\sum_{i=1}^{3} b_i \varphi_3^{(i)} + \alpha\, \chi_{\rho,\rho}\}\, da.$$

It follows from (5.129) and (5.130) that

$$\hat{u}_\alpha = -\frac{1}{24} b_\alpha x_3^4 - \frac{1}{6} c_\alpha x_3^3 - \frac{1}{2} d_\alpha x_3^2 - \frac{1}{2} c_4\, \varepsilon_{\alpha\beta} x_\beta x_3^2 -$$

$$- d_4 \varepsilon_{\alpha\beta} x_\beta x_3 + \sum_{i=1}^{3} (d_j + c_j x_3 + \frac{1}{2} b_j x_3^2) u_\alpha^{(i)} + w_\alpha\ ,$$

$$\hat{u}_3 = \frac{1}{6}(b_\rho x_\rho + b_3) x_3^3 + \frac{1}{2}(c_\rho x_\rho + c_3) x_3^2 + (d_\rho x_\rho + d_3) x_3 +$$

$$+ (c_4 x_3 + d_4)\varphi + x_3 \chi + \Psi\ , \qquad (5.133)$$

$$\hat{\varphi}_\alpha = \varepsilon_{\alpha\beta}(\frac{1}{6} b_\beta x_3^3 + \frac{1}{2} c_\beta x_3^2 + d_\beta x_3) + (c_4 x_3 + d_4)\psi_\alpha +$$

$$+ x_3 \chi_\alpha + \Psi_\alpha\ ,$$

$$\hat{\varphi}_3 = \sum_{i=1}^{3} (\frac{1}{2} b_i x_3^2 + c_i x_3 + d_i)\varphi_3^{(i)} + \frac{1}{2} c_4 x_3^2 + d_4 x_3 + w_3\ ,$$

where $u^\circ = (w_1, w_2, \Psi, \Psi_1, \Psi_2, w_3)$. The constitutive equations imply that

$$t_{\alpha\beta}(\hat{u}) = \lambda[\frac{1}{2}(b_\rho x_\rho + b_3) x_3^2 + (c_\rho x_\rho + c_3) x_3 + d_\rho x_\rho + d_3]\delta_{\alpha\beta} +$$

$$+ \lambda(\chi + c_4 \varphi)\delta_{\alpha\beta} + \sum_{i=1}^{3} (\frac{1}{2} b_i x_3^2 + c_i x_3 + d_i) T_{\alpha\beta}(w^{(i)}) + T_{\alpha\beta}(\omega^\circ),$$

$$t_{33}(\hat{u})=(\lambda+2\mu+\varkappa)[d_\rho x_\rho+d_3+(c_\rho x_\rho+c_3)x_3+\frac{1}{2}(b_\rho x_\rho+b_3)x_3^2]+$$
$$+(\lambda+2\mu+\varkappa)(\chi+c_4\varphi)+\lambda\sum_{i=1}^{3}(\frac{1}{2}b_i x^2+c_i x_3+d_i)u^{(i)}_{\rho,\rho}+\lambda w_{\alpha,\alpha},$$

$$t_{\alpha 3}(\hat{u})=P_\alpha(\omega)+x_3 P_\alpha(u')+(d_4+c_4 x_3)[P_\alpha(\hat{w})+\mu\varepsilon_{\beta\alpha}x_\beta]+$$
$$+\mu\sum_{i=1}^{3}(c_i+b_i x_3)u^{(i)}_\alpha, \qquad (5.134)$$

$$t_{3\alpha}(\hat{u})=Q_\alpha(\omega)+x_3 Q_\alpha(u')+(d_4+c_4 x_3)[Q_\alpha(\hat{w})+(\mu+\varkappa)\varepsilon_{\beta\alpha}x_\beta]+$$
$$+(\mu+\varkappa)\sum_{i=1}^{3}(c_i+b_i x_3)u^{(i)}_\alpha,$$

$$m_{\lambda\nu}(\hat{u})=H_{\lambda\nu}(\omega)+x_3 H_{\lambda\nu}(u')+(d_4+c_4 x_3)[H_{\lambda\nu}(\hat{w})+\delta_{\lambda\nu}]+$$
$$+\alpha\delta_{\lambda\nu}\sum_{i=1}^{3}(c_i+b_i x_3)\varphi_3^{(i)},$$

$$m_{33}(\hat{u})=(\alpha+\beta+\gamma)[d_4+c_4 x_3+\sum_{i=1}^{3}(c_i+b_i x_3)\varphi_3^{(i)}]+$$
$$+\alpha(d_4+c_4 x_3)\psi_{\rho,\rho}+\alpha(\Psi_{\rho,\rho}+x_3\chi_{\rho,\rho}),$$

$$m_{\alpha 3}(\hat{u})=\beta\varepsilon_{\alpha\nu}(d_\nu+c_\nu x_3+\frac{1}{2}b_\nu x_3^2)+\beta(\chi_\alpha+c_4\psi_\alpha)+$$
$$+\sum_{i=1}^{3}(d_i+c_i x_3+\frac{1}{2}b_i x_3^2)M_{\alpha 3}(w^{(i)})+M_{\alpha 3}(\omega^\circ),$$

$$m_{3\alpha}(\hat{u})=\gamma\varepsilon_{\alpha\nu}(d_\nu+c_\nu x_3+\frac{1}{2}b_\nu x_3^2)+\gamma(\chi_\alpha+c_4\psi_\alpha)+$$
$$+\beta\sum_{i=1}^{3}(d_i+c_i x_3+\frac{1}{2}b_i x_3^2)\varphi^{(i)}_{3,\alpha}+\beta w_{3,\alpha},$$

where $\omega^0 = (w_1, w_2, 0, 0, 0, w_3)$, $\omega = (0, 0, \Psi, \Psi_1, \Psi_2, 0)$.

The equations of equilibrium and the conditions on the lateral boundary reduce to

$$(T_{\beta\alpha}(\omega^0))_{,\beta} + h_\alpha = 0,$$

$$(M_{\alpha 3}(\omega^0))_{,\alpha} + \varepsilon_{\alpha\beta} T_{\alpha\beta}(\omega^0) + g = 0 \quad \text{on } \Sigma , \tag{5.135}$$

$$T_{\beta\alpha}(\omega^0) n_\beta = p^0_\alpha \quad , \quad M_{\alpha 3}(\omega^0) n_\alpha = q^0 \quad \text{on } \Gamma ,$$

and

$$L_i \omega = \gamma_i \quad \text{on } \Sigma , \quad N_i \omega = \rho_i \quad \text{on } \Gamma , \tag{5.136}$$

where

$$h_\alpha = \lambda(\chi + c_4 \varphi)_{,\alpha} + Q_\alpha(u') + c_4 [Q_\alpha(u') + (\mu + \varkappa)\varepsilon_{\beta\alpha} x_\beta] + + (\mu + \varkappa) \sum_{i=1}^{3} b_i w_\alpha^{(i)} + f_\alpha ,$$

$$g = \beta(\chi_\alpha + c_4 \psi_\alpha)_{,\alpha} + (\alpha + \beta + \gamma)(c_4 + \sum_{i=1}^{3} b_i \varphi_3^{(i)}) + + \alpha(\chi_\rho + c_4 \psi_\rho)_{,\rho} + g_3 ,$$

$$p^0_\alpha = p_\alpha - \lambda(\chi + c_4 \varphi) n_\alpha , \quad q^0 = k_3 - \beta(\chi_\alpha + c_4 \psi_\alpha) n_\alpha , \tag{5.137}$$

$$\gamma_\nu = - \sum_{i=1}^{3} c_i [\alpha \varphi^{(i)}_{3,\nu} + \beta \varphi^{(i)}_{3,\nu} - \varkappa \varepsilon_{\nu\beta} u^{(i)}_\beta] - \gamma \varepsilon_{\nu\beta} c_\beta - g_\nu ,$$

$$\gamma_3 = -(\lambda + \mu) \sum_{i=1}^{3} c_i w^{(i)}_{\alpha,\alpha} - (\lambda + 2\mu + \varkappa)(c_\rho x_\rho + c_3) - f_3 ,$$

$$\rho_\nu = k_\nu - n_\nu \alpha \sum_{i=1}^{3} c_i \varphi_3^{(i)}, \quad \rho_3 = p_3 - \mu \sum_{i=1}^{3} c_i w_\alpha^{(i)} n_\alpha .$$

It follows from (5.130),(5.132),(5.137), the divergence theorem and Theorem 5.12 that

$$\int_\Sigma h_\alpha da + \int_\Gamma p_\alpha^0 ds = G_\alpha - R_\alpha(\hat{u}_{,3}) = G_\alpha - \varepsilon_{\beta\alpha} H_\beta(\hat{u}_{,33}) = G_\alpha + D_{\alpha i} b_i = 0 \ ,$$

$$\int_\Sigma (\varepsilon_{\alpha\beta} x_\alpha h_\beta + g) da + \int_\Gamma (\varepsilon_{\alpha\beta} x_\alpha p_\beta^0 + q^0) ds = Z_3 - H_3(\hat{u}_{,3}) = 0 \ ,$$

$$\int_\Sigma \gamma_3 da - \int_\Gamma \rho_3 ds = -\int_\Sigma f_3 da - \int_\Gamma p_3 ds - D_{3j}\ c_j = 0 \ .$$

Thus we conclude that the necessary and sufficient conditions to solve the boundary-value problems (5.135) and (5.136) are satisfied.

From (5.16),(5.134) and (5.132) we find

$$H_\alpha(\hat{u}_{,3}) = \varepsilon_{\beta\alpha} D_{\beta i} c_i = \int_\Gamma (\varepsilon_{\alpha\beta} x_\beta p_3 + k_\alpha) ds + \int_\Sigma (\varepsilon_{\alpha\beta} x_\beta f_3 + g_\alpha) da + \varepsilon_{\alpha\beta} F_\beta ,$$

and by Theorem 5.12,

$$H_\alpha(\hat{u}_{,3}) = \int_\Gamma (\varepsilon_{\alpha\beta} x_\beta p_3 + k_\alpha) ds + \int_\Sigma (\varepsilon_{\alpha\beta} x_\beta f_3 + g_\alpha) da + \varepsilon_{\alpha\beta} R_\beta(\hat{u}).$$

The last two relations imply that $R_\alpha(\hat{u}) = F_\alpha$.

The conditions $R_3(\hat{u}) = F_3$ and $\underline{H}(\hat{u}) = \underline{M}$ reduce to

$$D_{ij}\ d_j = r_i \ ,$$

$$Dd_4 = -M_3 - \int_\Sigma \Big\{ \varepsilon_{\alpha\beta} x_\alpha \Big[\mu \Psi_{,\beta} + \varkappa \varepsilon_{\nu\beta} \Psi_{,\nu} + (\mu + \varkappa) \sum_{i=1}^{3} c_i w_\beta^{(i)} \Big] + \tag{5.138}$$

$$+ (\alpha + \beta + \gamma) \sum_{i=1}^{3} c_i \varphi_3^{(i)} + \alpha \Psi_{,\nu\nu} \Big\} da \ ,$$

where

$$r_\alpha = \varepsilon_{\alpha\beta}M_\beta - \int_\Sigma \{x_\alpha[\lambda w_{\rho,\rho} + (\lambda+2\mu+\varkappa)(\chi + c_4\varphi)] -$$
$$- \varepsilon_{\alpha\beta}[\gamma(\chi_{,\beta} + c_4\psi_\beta) + \beta w_{3,\beta}]\}da,$$

$$r_3 = -F_3 - \int_\Sigma [(\lambda+2\mu+\varkappa)(\chi + c_4\varphi) + \lambda w_{\alpha,\alpha}]da .$$

Thus, the vector $\hat{d}$ is defined by (5.138). □

Let us study now the Almansi problem. Let u^* be an equilibrium vector field on B which corresponds to the body loads $\{\underline{f} = \underline{f}^* x_3^n ,\ \underline{g} = \underline{g}^* x_3^n\}$, and satisfies the conditions

$$\underline{s}(u^*) = \underline{p}^* x_3^n ,\quad \underline{q}(u^*) = \underline{k}^* x_3^n \text{ on } \Pi ,\quad \underline{R}(u^*) = \underline{0} ,\quad \underline{H}(u^*) = \underline{0} , \tag{5.139}$$

where $\underline{f}^*, \underline{g}^*, \underline{p}^*$ and $\underline{k}^*$ are prescribed vector fields independent of x_3, and n is a positive integer or zero. Let u be an equilibrium field on B which corresponds to the body loads $\{\underline{f} = \underline{f}^* x_3^{n+1},\ \underline{g} = \underline{g}^* x_3^{n+1}\}$ and satisfies the conditions

$$\underline{s}(u) = \underline{p}^* x_3^{n+1} ,\quad \underline{q}(u) = \underline{k}^* x_3^{n+1} \text{ on } \Pi ,\quad \underline{R}(u) = \underline{0} ,\quad \underline{H}(u) = \underline{0}. \tag{5.140}$$

As in Section 2.3 we can prove that Almansi's problem reduces to the finding a vector field u once the vector field u^* is known. Moreover, we are led to seek the vector field u in the form

$$u = (n+1)[\int_0^{x_3} u^* dx_3 + v\{\hat{a}\} + w] , \tag{5.141}$$

where $\hat{a} = (a_1, a_2, a_3, a_4)$ is an unknown four-dimensional vector and w is an unknown vector field independent of x_3.
By (5.141) and the constitutive equations, we have

$$\begin{aligned} t_{ij}(u) &= (n+1)[\int_0^{x_3} t_{ij}(u^*)dx_3 + \sum_{r=1}^{4} a_r t_{ij}(v^{(r)}) + t_{ij}(w) + k_{ij}], \\ m_{ij}(u) &= (n+1)[\int_0^{x_3} m_{ij}(u^*)dx_3 + \sum_{r=1}^{4} a_r m_{ij}(v^{(r)}) + m_{ij}(w) + b_{ij}], \end{aligned} \tag{5.142}$$

where

$$k_{\alpha\beta} = \lambda\delta_{\alpha\beta}u_3^*(x_1,x_2,0), \quad k_{33} = (\lambda+2\mu+\varkappa)u_3^*(x_1,x_2,0) ,$$

$$k_{\alpha 3} = \mu u_\alpha^*(x_1,x_2,0) , \quad k_{3\alpha} = (\mu+\varkappa)u_\alpha^*(x_1,x_2,0) ,$$

$$h_{\eta\nu} = \alpha\delta_{\eta\nu}\varphi_3^*(x_1,x_2,0), \ h_{33} = (\alpha+\beta+\gamma)\varphi_3^*(x_1,x_2,0),$$

$$h_{\alpha 3} = \beta\varphi_\alpha^*(x_1,x_2,0) , \quad h_{3\alpha} = \gamma\varphi_\alpha^*(x_1,x_2,0) .$$

The equations of equilibrium and the conditions on the lateral boundary reduce to

$$(T_{\beta\alpha}(\omega))_{,\beta} + E_\alpha = 0,$$

$$(M_{\rho 3}(\omega))_{,\rho} + \varepsilon_{\alpha\beta}T_{\alpha\beta}(\omega) + J = 0 \quad \text{on } \Sigma, \tag{5.143}$$

$$T_{\beta\alpha}(\omega)n_\beta = p'_\alpha \ , \ M_{\alpha 3}(\omega)n_\alpha = q' \ \text{on } \Gamma,$$

and

$$L_i\omega^* = \zeta_i \ \text{on } \Sigma, \quad N_i\omega^* = \xi_i \ \text{on } \Gamma , \tag{5.144}$$

where

$$w = (v_1,v_2,v_3,\chi_1,\chi_2,\chi_3), \quad \omega = (v_1,v_2,0,0,0,\chi_3) ,$$

$$\omega^* = (0,0,v_3,\chi_1,\chi_2,0), \ E_\alpha = k_{\rho\alpha,\rho} + [t_{3\alpha}(u^*)](x_1,x_2,0),$$

$$J = h_{\alpha 3,\alpha} + [m_{33}(u^*)](x_1,x_2,0), \ p'_\alpha = -k_{\rho\alpha}n_\rho, \ q' = -h_{\rho 3}n_\rho, \tag{5.145}$$

$$\zeta_\alpha = -h_{\rho\alpha,\rho} - [m_{3\alpha}(u^*)](x_1,x_2,0),$$

$$\zeta_3 = -k_{\rho 3,\rho} + [t_{33}(u^*)](x_1,x_2,0),$$

$$\xi_\alpha = -h_{\rho\alpha}n_\rho , \ \xi_3 = -k_{\rho 3}n_\rho .$$

It follows from (5.145) that

$$\int_{\Sigma} E_{\alpha} da + \int_{\Gamma} p'_{\alpha} ds = - R_{\alpha}(u^*) = 0,$$

$$\int_{\Sigma} (\varepsilon_{\alpha\beta} x_{\alpha} E_{\beta} + J) da + \int_{\Gamma} (\varepsilon_{\alpha\beta} x_{\alpha} p'_{\beta} + q') ds = - H_3(u^*) = 0,$$

$$\int_{\Sigma} \zeta_3 da - \int_{\Gamma} \xi_3 ds = -R_3(u^*) = 0 .$$

We conclude that the necessary and sufficient conditions to solve the boundary value problems (5.143) and (5.144) are satisfied. We shall assume that the functions v_i and χ_i are known.

By Theorem 5.12, $R_{\alpha}(u) = \varepsilon_{\beta\alpha} H_{\beta}[(n+1)u^*] = 0$.

The conditions $R_3(u) = 0$, $\underline{H}(u) = \underline{0}$ reduce to

$$D_{\alpha j} a_j = - \int_{\Sigma} [x_{\alpha}(k_{33} + t_{33}(w)) - \varepsilon_{\alpha\rho}(b_{3\rho} + m_{3\rho}(w))] da,$$

$$D_{3j} a_j = - \int_{\Sigma} [k_{33} + t_{33}(w)] da ,$$

$$D a_4 = - \int_{\Sigma} \{\varepsilon_{\alpha\beta} x_{\alpha} [k_{3\beta} + t_{3\beta}(w)] + m_{33}(w) + b_{33}\} da .$$

This system determines the constants a_1, a_2, a_3 and a_4.

The problems of Almansi and Michell for Cosserat elastic bodies have been studied in the papers [131],[22],[71] by the semi-inverse method.

5.11. Applications

In this section we present some results concerning the relaxed Saint-Venant's problem for a circular cylinder. We assume that Γ is a circle of radius a and that the x_3-axis passes through the center of the cross section.

(i) Bending. The solution has the form (5.89) where $\{u_\alpha^{(\rho)}, \varphi_3^{(\rho)}\}$ ($\rho = 1,2$) are characterized by the plane strain problems (5.74), (5.75). Using the polar coordinates we obtain

$$u_r^{(1)} = [1 + \frac{\varkappa}{(2\mu+\varkappa)(1-3c)}]A_1 r^2 \cos\theta + \frac{\varkappa}{2(\mu+\varkappa)\delta}[I_2(\delta r) - I_0(\delta r)]A_2 \cos\theta ,$$

$$u_\theta^{(1)} = -\frac{1}{1-3c}(3 - c + \frac{3}{2\mu+\varkappa})A_1 r^2 \sin\theta + \frac{\varkappa}{2\delta(\mu+\varkappa)}[I_2(\delta r) + I_0(\delta r)]A_2 \cos\theta ,$$

$$\varphi_3^{(1)} = A_2 I_1(\delta r)\sin\theta - \frac{8(\mu+\varkappa)}{1-3c}A_1 r \sin\theta ,$$

where

$$A_1 = -\delta^2(1-3c)(8\varkappa h^o)^{-1}(\beta+\gamma\nu),$$

$$h^o = [4(\mu+\varkappa)-2c(2\mu+\varkappa)][8\varkappa I_2(\delta a)]^{-1}\delta^2 a^2 I_1'(\delta a) + 1 ,$$

$$A_2 = \delta a^2(\beta+\gamma\nu)[8\varkappa\gamma h^o I_2(\delta a)]^{-1}[4(\mu+\varkappa) - 2c(2\mu+\varkappa)],$$

$$\delta^2 = \frac{\varkappa(2\mu+\varkappa)}{\gamma(\mu+\varkappa)}, \quad c = \frac{\mu+\varkappa}{\lambda+2\mu+\varkappa}, \quad \nu = \frac{\lambda}{2\lambda+2\mu+\varkappa}.$$

Here and in what follows we denote by I_n and K_n the modified Bessel functions of order n. The other plane strain problem has the solution

$$u_r^{(2)} = [1 + \frac{\varkappa}{(2\mu+\varkappa)(1-3c)}]B_1 r^2 \sin\theta - \frac{\varkappa}{2\delta(\mu+\varkappa)}[I_2(\delta r) - I_0(\delta r)]B_2 \sin\theta ,$$

$$u_\theta^{(2)} = \frac{1}{1-3c}(3-c+\frac{3\varkappa}{2\mu+\varkappa})B_1 r^2 \cos\theta + \frac{\varkappa}{2\delta(\mu+\varkappa)}[I_2(\delta r) + I_0(\delta r)]B_2 \cos\theta ,$$

$$\varphi_3^{(2)} = B_2\, I_1(\delta r)\cos\theta + \frac{8(\mu+\varkappa)}{(1-3c)(2\mu+\varkappa)} B_1\, r\cos\theta \;,$$

where

$$B_1 = -\,\delta^2(1-3c)(\beta+\gamma\nu)(8\varkappa h^0)^{-1} \;,$$

$$B_2 = -(\beta+\gamma\nu)\,\delta\, a^2[8\varkappa h^0\gamma\, I_2(\delta a)]^{-1}[4(\mu+\varkappa)-2c(2\mu+\varkappa)].$$

It follows from (5.84) that

$$D_{11} = D_{22} = \frac{\pi a^4}{4} E + \pi a^2(\beta\nu+\gamma) - \pi a^2\, c\beta\varkappa\gamma^{-1}\delta^{-2} +$$

$$+ \frac{1}{4}\pi a^4\, c\lambda C - 2\pi a\, Q^{-1}\, I_1(\delta a)\beta \;,$$

$$D_{12} = 0,$$

where

$$E = (2\mu+\varkappa)(3\lambda+2\mu+\varkappa)(2\lambda+2\mu+\varkappa)^{-1} \;,$$

$$C = 8\gamma\delta\nu(\lambda c\, a^2\, Q)^{-1}\, I_2(\delta a) \;,$$

$$Q = \gamma\delta(\beta+\gamma\nu)^{-1}[I_2(\delta a) + I_0(\delta a) +$$

$$+ 8\varkappa\nu I_2(\delta a)(\lambda c\, a^2\delta^2)^{-1}].$$

Thus, we obtain

$$a_\alpha = \frac{1}{D_{11}}\varepsilon_{\alpha\beta}M_\beta \;.$$

The solution of this problem has been established in [88].

ii) <u>Torsion.</u> We seek the torsion functions in the form

$$\varphi = 0 \;, \quad \psi_\alpha = x_\alpha \Psi(r) \;,$$

where Ψ is an unknown function and $r^2 = x_\alpha x_\alpha$. The equations for the torsion functions reduce to

$$\Psi'' + \frac{3}{r}\Psi' - s^2\Psi = 0 \quad , \tag{5.146}$$

where

$$\Psi' = \frac{d\Psi}{dr} \quad , \quad s^2 = \frac{2}{\alpha+\beta+\gamma} \; .$$

We introduce the function F by $F = r\Psi$. Then the equation (5.146) becomes

$$F'' + \frac{1}{r}F' - (\frac{1}{r^2} + s^2)F = 0.$$

The solution of this equation is

$$F = B_1^0\, I_1(sr) + B_2^0\, K_1(sr) \; ,$$

where B_1^0 and B_2^0 are arbitrary constants. In order to obtain a solution which is finite in $r = 0$ we take $B_2^0 = 0$. Thus, we have

$$u_\alpha = \varepsilon_{\beta\alpha} a_4 x_\beta x_3 \; , \quad u_3 = 0 \; ,$$

$$\varphi_\alpha = a_4[\frac{1}{r}\, B_1^0\, I_1(sr) - \frac{1}{2}]x_\alpha, \quad \varphi_3 = 0.$$

The conditions on boundary of Σ reduce to

$$a(\alpha+\beta+\gamma)F'(a) + \alpha F(a) = \frac{1}{2}(\beta+\gamma) \; . \tag{5.147}$$

If we take into account the relation $I_1' + I_1 = I_0$, it follows from (5.147) that

$$B_1^0 = a\, s\, I_0(as)[I_1(as)]^{-1} - (\beta+\gamma)(\alpha+\beta+\gamma)^{-1}.$$

The torsional rigidity is given by

$$D = \frac{1}{4}\pi a^4(2\mu+\varkappa) + \pi a^2(\beta+\gamma) + 2\pi\varkappa B_1^0\int_0^a x^2 I_1(sx)dx \; +$$

$$+ \; 2\pi B_1^0\alpha\int_0^a [sxI_1'(sx) + I_1(sx)]dx \; .$$

With the help of the relations

$$[x^2 I_2(x)]' = x^2 I_1(x), \quad x I_1'(x) + I_1(x) = [x I_1(x)]',$$
$$I_1(0) = 0,$$

we obtain

$$D = \frac{1}{4}\pi a^4(2\mu+\varkappa)+\pi a^2(\beta+\gamma) + \frac{2}{3}\pi\varkappa a^2 B_1^0 I_2(as)+2\pi a\alpha B_1^0 I_1(as).$$

The constant a_4 is determined by (5.83).

The torsion problem for a circular cylinder was studied in [153], [88]. The flexure problem has been solved in [89],[90].

REFERENCES

1 E.Almansi, Sopra la deformazione dei cilindri solecitati lateralmente. Atti Accad.Naz.Lincei Rend.Cl.Sci.Fis.Mat.Natur.Ser. 5, 10(1901), I: 333-338; II: 400-408.

2 G.L.Anderson, On forced vibration in the linear theory of micropolar elasticity. Int.J.Engng.Sci. 11(1973), 21-40.

3 N.H.Arutyunyan and B.L.Abramyan, Torsion of Elastic Bodies (in Russian). Fizmatghiz, Moscow, 1963.

4 Bai Zhe Zheng, On one aspect of Saint-Venant's principle. J.Huazhong Univ.Sci.Techn. 4(1982), 71-81.

5 R.Batra, Saint-Venant's principle in linear elasticity with microstructure. J.Elasticity 13(1983), 165-173.

6 E.Benvenuto, A.Campanella and G.M.Gancia, The elastic cylinder: general solution by means of polynomial expansion (in Italian). Atti Accad.Ligure Sci.Lett. 38(1982), 240-270.

7 V.L.Berdicevski, On the proof of the Saint-Venant principle for bodies of arbitrary shape (in Russian). Prikl.Mat.Meh.38(1974), 851-864.

8 C.A.Berg, The physical meaning of astatic equilibrium in Saint-Venant's principle for linear elasticity. J.Appl.Mech. 36(1969), 392-396.

9 K.Berglund, Generalization of Saint-Venant's principle to micropolar continua. Arch.Rational Mech.Anal. 64(1977), 317-326.

10 Y.Biollay, First boundary value problem in elasticity: bounds for the displacements in Saint-Venant's principle. ZAMP 31(1980), 556-567.

11 B.A.Boley, On a dynamical Saint-Venant principle. J.Appl.Mech. 27(1960), 74-78.

12 C.I.Borş, Theory of elasticity for anisotropic bodies (in Romanian). Editura Academiei, Bucureşti, 1970.

13 J.Boussinesq, Application des potentiels à l'étude de l'équilibre et des mouvements des solides élastiques. Gauthier-Villars, Paris, 1885.

14 S.Breuer and J.J.Roseman, On Saint-Venant's principle in three-dimensional nonlinear elasticity. Arch.Rational Mech.Anal. 63(1977), 191-203.

15 O.Brulin and R.K.T.Hsieh, Mechanics of Micropolar Media, World Scientific, Singapore, 1982.

16 D.E.Carlson, Linear Thermoelasticity. In vol.VI a/2 of the Handbuch der Physik, edited by C.Truesdell, Springer-Verlag, Berlin-Heidelberg-New York, 1972.

17 Chao Hwei Yuan, On the torsion of non-homogeneous anisotropic elastic cylinders. Sci.Sinica, 9(1960), 47-61.

18 S.Chiriţă, Thermal stresses in anisotropic cylinders. Acta Mechanica 23(1975), 301-306.

19 S.Chiriţă, Saint-Venant's problem for nonlocal elastic solids. An.St.Univ."Al.I.Cuza" Iaşi, s.Matematică, 24(1978), 147-156.

20 S.Chiriţă, Saint-Venant's problem for anisotropic circular cylinder. Acta Mechanica 34(1979), 243-250.

21 S.Chiriţă, Saint-Venant's problem for composite micropolar elastic cylinders. Int.J.Engng.Sci. 17(1979), 573-580.

22 S.Chiriţă, Deformation of loaded micropolar elastic cylinders. Int.J.Engng.Sci. 19(1981), 845-853.

23 I.Choi and C.O.Horgan, Saint-Venant's principle and end effects in anisotropic elasticity. J.Appl.Mech. Trans.ASME, Serie E, 44(1977), 424-430.

24 A.Clebsch, Theorie der Elasticität fester Körper, B.G.Teubner, Leipzig, 1862.

25 W.A.Day, Generalized torsion: The solution of a problem of Truesdell's.Arch.Rational Mech.Anal. 76(1981), 283-288.

26 E.Deutsch, On the centre of flexure. ZAMP 12(1961), 212-218.

27 J.B.Diaz and L.E. Payne, Mean value theorems in the theory of elasticity. Proc.Third.U.S.Natt.Congress Appl.Mech.(1958), 293-303.

28 G.Y.Djanelidze, Saint-Venant's principle (in Russian), Leningrad. Politehn.Inst.Trudy. 192(1958).

29 G.Y.Djanelidze, Almansi problem. Leningrad Politehn.Inst.Trudy. 210(1960), 25-38.

30 J.L.Ericksen, Equilibrium of bars. J.Elasticity 5(1975), 191-201.

31 J.L.Ericksen, On the formulation of St.Venant's problem. Nonlinear Analysis and Mechanics, Heriot-Watt Symposium, vol.1, 158-186, Pitman, London, 1977.

32 J.L.Ericksen, On St.Venant's problem for thin-walled tubes. Arch.Rational Mech.Anal. 70(1979), 7-12.

33 J.L.Ericksen, On the status of St.Venant's solutions as minimizers of energy. Int.J.Solids and Structures 16(1980), 195-198.

34 A.C.Eringen and E.S.Suhubi, Nonlinear theory of simple microelastic solids. Int.J.Engng.Sci. 2(1964), I:189-203, II:389-404.

35 A.C.Eringen, Theory of Micropolar Elasticity. Fracture vol.2, Academic Press, New York, 1968.

36 G.Fichera, Sulla torsione elastica dei prismi cavi. Rend.Mat. Ser.5, 12(1953), 163-176.

37 G.Fichera, Existence Theorems in Elasticity. In vol.VI a/2 of the Handbuch der Physik, edited by C.Truesdell, Springer-Verlag, Berlin-Heidelberg-New York, 1972.

38 G.Fichera, Il principio di Saint-Venant: intuizione dell'ingegnere e rigore del matematico. Rend.Mat., Ser.VI, 10(1977), 1-24.

39 G.Fichera, Remarks on Saint-Venant's principle. Complex Analysis and its Applications. I.N.Vekua Anniversary Volume, 543-554, Moscow, 1978.

40 G.Fichera, Problemi Analitici Nuovi nella Fisica Matematica Classica. Quaderni del Consiglio Nazionale delle Ricerche, Gruppo Nazionale di Fisica Matematica, Scuola Tipo-Lito "Instituto Anselmi", Napoli, 1985.

41 J.N.Flavin, Another aspect of Saint-Venant's principle in elasticity. ZAMP 29(1978), 328-332.

42 R.D.Gauthier and W.E.Jahsman, Bending of a curved bar of micropolar elastic material. J.Appl.Mech. 43(1976), 502-503.

43 T.G.Gegelia and R.K.Chichinadze, Basic static problems of elastic micropolar media. Arch.Mech. 28(1976), 89-104.

44 A.E.Green and R.S.Rivlin, Multipolar continuum mechanics. Arch. Rational Mech.Anal. 17(1964), 113-147.

45 A.E.Green, Micro-materials and multipolar continuum mechanics. Int.J.Engng.Sci. 3(1965), 533-537.

46 G.Grioli, Mathematical Theory of Elastic Equilibrium. Springer-Verlag, Berlin-Göttingen-Heidelberg, 1962.

47 M.E.Gurtin, The Linear Theory of Elasticity. In vol.VI a/2 of the Handbuch der Physik, edited by C.Truesdell, Springer-Verlag, Berlin,Heidelberg,New York, 1972.

48 G.M.Hatiashvili, The problem of Almansi for a composite anisotropic cylindrical body (in Russian). Trudy Vcisl.Tsentra.Akad. Nauk Gruzin. SSR, 4(1963), 29-42.

49 G.M.Hatiashvili, Almansi-Michell Problems for Homogeneous and Composed Bodies (in Russian), Izd.Metznierreba, Tbilisi, 1983.

50 I.Hlavacek and M.Hlavacek, On the existence and uniqueness of solution and some variational principles in linear theories of elasticity with couple-stresses. Aplikace Matematiky, 14(1969), I: 387-410, II: 411-426.

51 C.O.Horgan and J.K.Knowles, The effect of nonlinearity on the principle of Saint-Venant. J.Elasticity, 11(1981), 271-292.

52 C.O.Horgan, Saint-Venant's principle in anisotropic elasticity theory. Mechanical behavior of anisotropic solids. Proc.Euromech.Colloq. 115. Villard-de-Lans / France 1979, Colloq.Int. CNRS 295, 853-868(1982).

53 C.O.Horgan and J.K.Knowles, Recent developments concerning Saint-Venant's principle. Adv.Appl.Mech. 23(1983), 179-269.

54 G.Horvay, Saint-Venant's principle: a biharmonic eigenvalue problem. J.Appl.Mech. 24(1957), 381-386.

55 D.Ieşan, Tensions thermiques dans des barres élastiques non-homogènes. Ann.St.Univ."Al.I.Cuza" Iaşi, s.Matematică, 14(1968), 177-203.

56 D.Ieşan, On the generalized plane strain (in Romanian). St.Cerc. Mat. 21(1969), 595-609.

57 D.Ieşan, Existence theorems in micropolar elastostatics. Int.J. Engng.Sci. 9(1971), 59-78.

58 D.Ieşan, Torsion of micropolar elastic beams. Int.J.Engng.Sci. 9(1971), 1047-1060.

59 D.Ieşan, On Saint-Venant's problem in micropolar elasticity. Int.J.Engng.Sci. 9(1971), 879-888.

60 D.Ieşan, Flexure of micropolar elastic beams. Rev.Roum.Math. Pures et Appl. 17(1972), 709-720.

61 D.Ieşan, On Almansi's problem for elastic cylinders. Atti.Accad. Sci.Ist.Bologna.Cl.Sci.Fis.Mat.Rend. s.XII, 9(1972), 128-139.

62 D.Ieşan, On the thermal stresses in beams. J.Eng.Math. 6(1972), 155-163.

63 D.Ieşan, On the second boundary value problem in the linear theory of micropolar elasticity. Atti Accad.Naz.Lincei Rend.Cl. Sci.Fis.Mat.Natur.s.VIII, 55(1973), 456-459.

64 D.Ieşan, Torsion of anisotropic micropolar elastic cylinders. ZAMM 54(1974), 773-779.

65 D.Ieşan, Thermal stresses in micropolar elastic cylinders. Acta Mechanica 21(1975), 261-272.

66 D.Ieşan, Saint-Venant's problem for heterogeneous elastic cylinders. Engng.Trans. 24(1976), 289-306.

67 D.Ieşan, Saint-Venant's problem for inhomogeneous and anisotropic elastic bodies. J.Elasticity, 6(1976), 277-294.

68 D.Ieşan, Saint-Venant's problem for heterogeneous anisotropic elastic solids. Ann.Mat.Pura Appl. (IV), 108(1976), 149-159.

69 D.Ieşan, Saint-Venant's problem for inhomogeneous bodies. Int.J. Engng.Sci. 14(1976), 353-360.

70 D.Ieşan, Saint-Venant's problem for inhomogeneous and anisotropic elastic solids with microstructure. Arch.Mech. 29(1977), 419-430.

71 D.Ieşan, Deformation of composed Cosserat elastic cylinders. Rev.Roum.Sci.Techn.Mec.Appl. 23(1978), 823-845.

72 D.Ieşan, Saint-Venant's problem for inhomogeneous and anisotropic Cosserat elastic solids. ZAMM 58(1978), 95-99.

73 D.Ieşan, Theory of Thermoelasticity (in Roumanian), Romanian Academy Publ. House, Bucharest, 1979.

74 D.Ieşan, Thermal stresses in composite cylinders. J.Thermal Stresses, 3(1980), 495-508.

75 D.Ieşan, Torsion problem in the theory of non-simple elastic materials. An.St.Univ. "Al.I.Cuza" Iaşi, s.Matematică, 27(1981), 167-172.

76 D.Ieşan, On generalized Saint-Venant's problems. Int.J.Engng. Sci. 24(1986), 849-858.

77 D.Ieşan, On the solution of a problem of Truesdell's. An.St. Univ."Al.I.Cuza" Iaşi, s.Matematică, 32(1986).

78 D.Ieşan, Generalized twist for the torsion of micropolar cylinders. Meccanica, 21(1986).

79 D.Ieşan, On Saint-Venant's problem. Arch.Rational Mech.Anal. 91(1986), 363-373.

80 J.Ignaczak, A dynamic version of Saint-Venant's principle in the linear theory of elasticity. Bull.Acad.Polon.Sci., Sér.Sci. Techn. 22(1974), 483-489.

81 H.B.Keller, Saint-Venant's procedure and Saint-Venant's principle. Quart.Appl.Math. 22(1965), 293-304.

82 R.J.Knops and L.E.Payne, Uniqueness theorems in linear elasticity. In: Springer Tracts in Natural Philosophy, edited by B.D.Coleman, vol.19, Springer-Verlag, Berlin,Heidelberg,New York, 1971.

83 R.J.Knops and L.E.Payne, A Saint-Venant principle for nonlinear elasticity. Arch.Rational Mech.Anal. 81(1983), 1-12.

84 J.K.Knowles, On Saint-Venant's principle in the two-dimensional linear theory of elasticity. Arch.Rational Mech.Anal.21(1966), 1-22.

85 J.K.Knowles and E.Sternberg, On Saint-Venant's principle and the torsion of solids of revolution. Arch.Rational Mech.Anal. 21(1966) 100-120.

86 J.K.Knowles and C.O.Horgan, On the exponential decay of stresses in circular elastic cylinders subject to axisymmetric self-equilibrated end loads. Int.J.Solids and Structures 5(1969), 33-50.

87 J.K.Knowles, An energy estimate for the biharmonic equation and its application to Saint-Venant's principle in plane elastostatics. Indian J.Pure Appl.Math. 14(1983), 791-805.

88 G.V.Krishna Reddy and N.K.Venkatasubramanian, Saint-Venant's problem for a micropolar elastic circular cylinder. Int.J.Engng.Sci. 14(1976), 1047-1057.

89 G.V.Krishna Reddy and N.K.Venkatasubramanian, Flexure of a micropolar elastic circular cylinder. Proc.Indian Acad.Sci. 86A(1977), 575-590.

90 G.V.Krishna Reddy and N.K.Venkatasubramanian, On the flexural rigidity of a micropolar elastic circular cylinder. J.Appl.Mech. 45(1978), 429-431.

91 S.A.Kuliev, F.G.Radzhabov, Torsion of prismatic beams reinforced by a round rod. Izv.Akad.Nauk Azerbaidzhan SSR Ser.Fiz.-Tekhn. 4(1983), 143-148.

92 V.D.Kupradze, T.G.Gegelia, M.O.Bashelishvili and T.V.Burchuladze, Three-Dimensional Problems of Mathematical Theory of Elasticity and Thermoelasticity, North-Holland Publ., 1979.

93 G.A.Kutateladze, On the torsion of hollow piecewise homogeneous beams (in Russian). Dakl.Akad.Nauk SSSR, 198(1971), 1295-1298.

94 P.Ladevèze, Principes de Saint-Venant en déplacement et en contrainte pour les poutres droités élastiques semi-infinies. ZAMP 33(1982), 132-139.

95 P.Ladevèze, Sur le principe de Saint-Venant en élasticité. J.Méc.Théor.Appl. 2(1983), 161-184.

96 S.G.Lekhnitskii, Theory of Elasticity of an Anisotropic Elastic Body, Holden-Day, Inc., San Francisco, 1963.

97 V.A.Lomakin, Theory of Nonhomogeneous Elastic Bodies (in Russian) MGU, Moscow, 1976.

98 A.E.H.Love, A Treatise on the Mathematical Theory of Elasticity, Fourth Edition, Cambridge University Press, 1934.

99 O.Maisonneuve, Sur le principe de Saint-Venant. Thèse. Université de Poitiers, 1971.

100 M.Matschinskii, Beweis des Saint-Venantschen Prinzips, ZAMM 39 (1959), 9-11.

101 S.I.Melnik, Estimates for the St.Venant principle (in Russian). Perm.Gos.Univ.Učen.Zap.Mat. 103(1963), 178-180.

102 J.H.Michell, The theory of uniformly loaded beams. Quart.J.Math. 32(1901), 28-42.

103 L.M.Milne-Thomson, Antiplane Elastic Systems. Springer-Verlag, Berlin-Göttingen-Heidelberg, 1962.

104 R.D.Mindlin, Microstructure in linear elasticity. Arch.Rational Mech.Anal. 16(1964), 51-77.

105 R.D.Mindlin, Solution of Saint-Venant's torsion problem by power series. Int.J.Solids and Structures 11(1975), 321-328.

106 v.R.Mises, On Saint-Venant's principle. Bull.Amer.Math.Soc. 51(1945), 555-562.

107 M.Mişicu, Torsion and Flexure (in Romanian). Editura Academiei, Bucharest, 1973.

108 R.G.Muncaster, Saint-Venant's problem in nonlinear elasticity: a study of cross-sections. Nonlinear Analysis and Mechanics; Heriot-Watt Symposium vol.IV, p.17-75, Pitman, London, 1979.

109 R.G.Muncaster, Saint-Venant's problem for slender prisms. Utilitas Math. 23(1983), 75-101.

110 N.I.Muskhelishvili, Sur le problème de torsion des poutres élastiques composées. C.R.Acad.Sci.Paris, 194(1932), 1435.

111 N.I.Muskhelishvili, Some Basic Problems of the Mathematical Theory of Elasticity. Noordhoff, Groningen, 1953.

112 P.M.Naghdi, The Theory of Shells and Plates. In vol.VI a/2 of the Handbuch der Physik, edited by C.Truesdell, Springer-Verlag, Berlin, Heidelberg, New York, 1972.

113 O.I.Napetvaridze, On the boundary value problems of the theory of elasticity with couple-stresses (in Russian). Seminar of Institute of Applied Mathematics. Abstract of papers. Tbilisi, 5(1971), 53-67.

114 M.Nicolesco, Les fonctions polyharmoniques. Actualités Scientifiques et Industrielles 331, Paris, 1936.

115 V.V.Novozhilov and L.I.Slepian, On St.Venant's principle in the dynamics of beams. J.Appl.Math.Mech. 29(1965), 293-315.

116 W.Nowacki, Theory of Asymmetric Elasticity. Polish Scientific Publishers, Warszawa and Pergamon Press, Oxford, New York, Paris, Frankfurt, 1986.

117 J.Nowinski and S.Turski, A study of states of stress in inhomogeneous bodies (in Polish), Arch.Mech.Stos. 5(1953), 397-414.

118 O.A.Oleinik and G.A.Yosifian, Boundary value problems for second order elliptic equations in unbounded domains and Saint-

Venant's principle. Ann.della Scuola Norm.Sup.Pisa, IV 2(1977), 269-290.

119 M.B.Orazov, Saint-Venant's principle for equations of steady oscillations of an elastic semicylinder. (in Russian). Izv. Akad.Nauk Turkmen.SSR Ser.Fiz.-Techn.Nauk. 5(1983), 3-8.

120 L.E.Payne, Bounds for solutions of a class of quasilinear elliptic boundary value problems in terms of the torsion function. Proc.Roy.Soc.Edinburgh Sect.A 88(1981), 251-265.

121 L.E.Payne and G.A.Philippin, Isoperimetric inequalities in the torsion and clamped membrane problems for convex plane domains. SIAM J.Math.Anal. 14(1983), 1154-1162.

122 L.E.Payne and L.T.Wheeler, On the cross section of minimum stress concentration in the Saint-Venant theory of torsion. J.Elasticity 14(1984), 15-18.

123 P.Podio-Guidugli, St.Venant formulae for generalized St.Venant problems. Arch.Rational Mech.Anal. 81(1983), 13-20.

124 A.Polania, On Saint-Venant's principle in three-dimensional elasticity. Meccanica, 11(1976), 98-101.

125 G.Polya and A.Weinstein, On the torsional rigidity of multiply connected cross-sections. Annals of Math. 52(1950), 154-163.

126 A.Radu, Saint-Venant's problem for nonhomogeneous bars. (in Romanian), An.St.Univ."Al.I.Cuza" Iaşi, Matematică, 12(1966), 415-430.

127 A.Robinson, Non-Standard Analysis. North Holland, Amsterdam, 1966.

128 J.J.Roseman, A pointwise estimate for the stress in a cylinder and its application to Saint-Venant's principle. Arch.Rational Mech.Anal. 21(1966), 23-48.

129 J.J.Roseman, The principle of Saint-Venant in linear and non-linear plane elasticity. Arch.Rational Mech.Anal. 26(1967), 142-162.

130 A.K.Rukhadze and R.T.Zivzivadze, A generalized Almansi problem for prismatic beams that are composed of various isotropic materials. Soobs.Akad.Nauk Gruzin.SSR 103(1981), 41-44.

131 G.Rusu, Theory of loaded Cosserat elastic cylinders. An.St.Univ. "Al.I.Cuza" Iaşi, s.Matematică, 24(1978), 379-388.

132 A.J.C.B. de Saint-Venant, Mémoire sur la torsion des prismes. Mémoires présentés par divers savants à l'Académie des Sciences. 14(1855), 233-560.

133 A.J.C.B. de Saint-Venant, Mémoire sur la flexion des prismes. Journal de Mathématiques de Liouville, Sér.II, 1(1856), 89-189.

134 V.S.Sarkisyan, On the method of the solution of elasticity theory problem for the inhomogeneous and anisotropic body. Differential equations and their applications, Equadiff. 5. Proc.5th Czech.Conf.Bratislava 1981, Teubner-Texte Math. 47(1982), 217-300.

135 R.D.Schile and R.L.Sierakowski, On the Saint-Venant problem for a non-homogeneous elastic material. Quart.Appl.Math. 23(1965), 19-25.

136 D.I.Sherman, On the problem of plane strain in non-homogeneous media. In "Nonhomogeneity in Elasticity and Plasticity", 3-12, Pergamon Press, London, 1959.

137 R.S.Shield, An energy method for certain second-order effect with application to torsion of elastic bars under torsion. J.Appl.Mech. 47(1980), 75-81.

138 R.T.Shield, Extension and torsion of elastic bars with initial twist. J.Appl.Mech., Trans.ASME 49(1982), 779-786.

139 I.S.Sokolnikoff, Mathematical Theory of Elasticity, Second Edition, McGraw-Hill, New York, 1956.

140 L.Solomon, Élasticité Linéaire. Masson, Paris, 1968.

141 E.Soós, On Saint-Venant's problem for a nonhomogeneous and anisotropic body (in Romanian). An.Univ.București Mat.Mec. 12(1973), 70-77.

142 E.Sternberg, On Saint-Venant's principle. Quart.Appl.Math., 11(1954), 393-402.

143 E.Sternberg and J.K.Knowles, Minimum energy characterizations of Saint-Venant's solution to the relaxed Saint-Venant problem. Arch.Rational Mech.Anal. 21(1966), 89-107.

144 E.Sternberg, On Saint-Venant torsion and the plane problem of elastostatics for multiply connected domains. Arch.Rational Mech.Anal. 85(1984), 295-310.

145 R.A.Toupin, Theories of elasticity with couple-stress. Arch. Rational Mech.Anal. 17(1964), 85-112.

146 R.A.Toupin, Saint-Venant's principle. Arch.Rational Mech.Anal. 18(1965), 83-96.

147 R.A.Toupin, Saint-Venant and a matter of principle. Trans.N.Y. Acad.Sci. 28(1965), 221-232.

148 C.Truesdell, The rational mechanics of materials - past, present, future. Appl.Mech.Reviews 12(1959), 75-80.

149 C.Truesdell, The rational mechanics of materials - past, present, future (Corrected and modified reprint of [148]), pp. 225-236 of Applied Mechanics Surveys, Spartan Books, 1966.

150 C.Truesdell, Some challenges offered to analysis by rational thermomechanics, pp.495-603 of Contemporary Developments in Continuum Mechanics and Partial Differential Equations, G.M. de la Penha & L.A.Medeiros, Eds., North-Holland, 1978.

151 C.Truesdell, History of Classical Mechanics. Die Naturwissenschaften 63, Part I: to 1800, 53-62; Part II: the 19th and 20th Centuries, 119-130. Springer-Verlag, 1976.

152 G.V.Tsikhistavi, Saint-Venant problem for homogeneous "curvilinearly conic" isotropic bodies with end loading. Trudy. Vychisl.Tsentra Akad.Nauk Gruzin. SSSR 22(1982), 123-131.

153 C.Usidus and M.Sokolovski, Torsion of Cylindrical Shafts Made of Micropolar Material. Bull.Acad.Polon.Sci., Sér.Sci.Techn. 21(1973), 19-23.

154 I.N.Vekua and A.K.Rukhadze, Torsion problem for a circular cylinder reinforced by a longitudinal circular rod. (in Russian). Izv.Akad.Nauk SSSR, 3(1933), 1297-1308.

155 W.Voigt, Theoretische Studien über die Elasticitätsverhältnisse der Krystalle. Abh.Ges.Wiss.Göttingen, 34(1887), 53-153.

156 O.Zanaboni, Dimostrazione generale del principio del De Saint-Venant. Atti Accad.Lincei Rend. 25(1937), 117-121.

157 O.Zanaboni, Valutazione dell'errore massimo cui dà luogo l'applicazione del principio del De Saint-Venant. Atti Accad.Lincei Rend. 25(1937), 595-601.

158 O.Zanaboni, Sull'approssimazione dovuta al principio del De Saint-Venant nei solidi prismatici isotropi. Atti Accad.Lincei Rend. 26(1937), 340-345.

159 L.M.Zubov, Theory of torsion of prismatic rods under finite deformations (in Russian). Dokl.Akad.Nauk SSSR 270(1983), 827-831.

160 C.Weber and W.Günther, Torsionstheorie. Braunschweig, Vieweg, 1958.

SUBJECT INDEX